IMAGES
of America

RANCHO SESPE

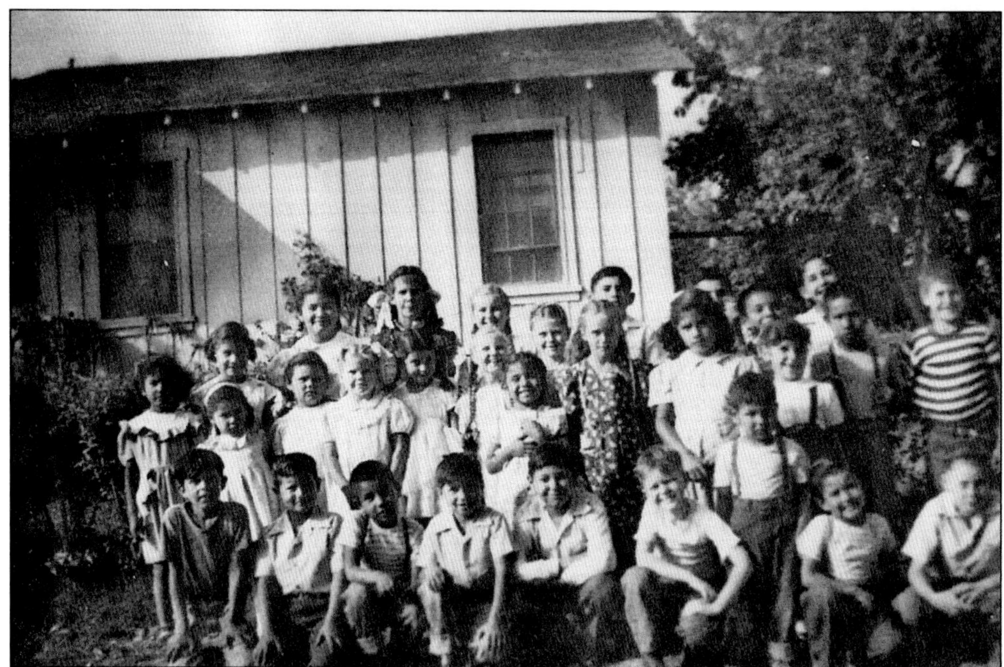

The children of Rancho Sespe celebrate Bobby Villanueva's birthday. He invited all his friends, and they posed for this photograph in front of Becky Sosa's and Gilbert Sosa Jr.'s house. Pictured are, from left to right, (first row) unidentified, Manuel Ramirez, Raymond Ramirez, unidentified, Robert "Bobby" Villanueva, Jimmy Pendergrass, James Torres, unidentified, and Tom Swift; (second row) Benny Torres, Angie Torres, Christina Ramirez, Patsy Campbell, May Ann Ramirez, unidentified, Linda Ramirez, Oma Crockett, Becky Sosa, and unidentified; (third row) Rachel Torres, Mary Lou Torres, Loretta Campbell, four unidentified, Tommy Ramirez, Gilbert Sosa, Arthus Ramirez, and Paul Casas. The children, who are now adults, recall the good times they had celebrating birthdays, playing together, and enjoying life as children, as they so often did when all together. (Courtesy of Robert Villanueva family collection.)

On the Cover: The cover shows sixth-grade students in front of Mountain View School with their teacher, Glen Phillips, in 1937. Some of these children were from Rancho Sespe; they were bused into Fillmore, where those who spoke Spanish attended Mountain View School on Mountain View Street and those who spoke English attended Sespe School on Sespe Avenue. Between 1925 and 1945, Mountain View had fifth- and sixth-grade classes for pupils of Mexican parentage. Segregation ended in 1945 at the end of World War II. Phillips went on to become an administrator in the Fillmore Unified School District, and he wrote many grants. One was the Early Childhood Education Grant, which provided learning materials and supplies as well as teacher training for kindergarten and first grade. Petra González is identified as the fifth girl from left. (Courtesy of Sosa-Morales family collection.)

IMAGES
of America

RANCHO SESPE

Becky Morales, Ernie Morales,
and Evie Ybarra

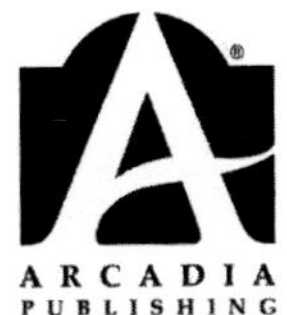

ISBN 978-1-4671-2496-6

Published by Arcadia Publishing
Charleston, South Carolina

Printed in the United States of America

Library of Congress Control Number: 2017944870

For all general information, please contact Arcadia Publishing:
Telephone 843-853-2070
Fax 843-853-0044
E-mail sales@arcadiapublishing.com
For customer service and orders:
Toll-Free 1-888-313-2665

Visit us on the Internet at www.arcadiapublishing.com

Rancho Sespe *is dedicated to all the families who lived and worked on the ranch from its early development to the present day.*

Becky Morales dedicates this book to her parents, Gilbert Lujan Sosa and Alice Margaret Sigala Sosa; to her big brother, Gilbert "Mo" Sosa; and to her aunt Glafida Sigala Rodarte.

Ernie Morales dedicates this book to his dad, Frank "Pancho" Morales, and to his mom, Georgia Morales. Their love of family provided their children—Frances, Ernie, and Sonia—with a keen sense of personal value and consideration of others.

Evie Ybarra dedicates this book to her grandmother Soledad Sanchez Ybarra, to her father, Marcelino "Woody" Ybarra, and to her mother, Minnie Ybarra. All set high standards for their family in an environment of integrity, generosity of spirit, kindness, and unconditional love.

CONTENTS

Foreword

The story of Rancho Sespe began in 1833, when the government of Mexico granted to Don Carlos Antonio Carrillo six square leagues of land along the Santa Clara River, from the arroyo of Piru on the east to the arroyo of Mupu on the west and from mountain top to mountain top. The stories in this book and the history of the ranch are really the stories of the people who owned the land and the families who lived and worked the land through the generations. It is through the efforts of Becky Sosa Morales, who grew up at Rancho Sespe; Ernie Morales; and author Evie Ybarra that these stories are now being told. The stories are of good times, family events, love, loss, and murder and have been researched exhaustively by Becky, Ernie, and Evie using historical records and extensive interviews and photographs of those who lived or grew up on the ranch. The Fillmore Historical Museum is pleased to have a room dedicated to the stories of the men, women, and families that owned and lived on the ranch. We also have an extensive collection of publications, records, and agricultural information collected from the ranch by historian Ynez Haase. We are indebted to all these individuals for their efforts in making this book possible.

—Martha Gentry
Executive Director, Fillmore Historical Museum

Acknowledgments

Thank-yous go to the many families who shared their personal stories and photographs for *Rancho Sespe* and who took the time to search through older photo albums and cupboards for all the vintage photographs. Many thanks go to the following people: Loretta Campbell Lombard; the Crockett and Renteria families; the Ramirez, González, Cowans, Pulido, Sosa, Torres, Rivera, and Rodarte families; Bobby Villanueva for his vintage photographs; Eva Leon Diaz, Henry Ramirez, Petra González, Dale Crockett, and Esther Robles; the Casas and Castorena families; Alex and Carmen Davis, Virgil Goines, and Gilbert Banuelos for their stories; and to Nancy Padelford Lesperance, who shared Martha Stoll's photographs. A thank-you goes to the Fillmore Historical Museum and director Martha Gentry, who shared many photographs of the early days of Rancho Sespe. These included rare photographs of Keith and Eudora Spalding. In addition, thanks go to Bob Crum for his special photographs of the Fillmore "F" on the hill. Dorothy Haase contributed a historical perspective about how the Fillmore Historical Museum was organized and formed for the city of Fillmore; many thanks, Dorothy.

To those who shared their stories, we are most grateful to you as well. You captured vivid memories of the past. Thanks go to Bobby Villanueva, Glafida Sigala Rodarte, Tomás Torres, Joe Pulido, Martha Stoll, Laura Castorena Escobar, George "Rudy" González, Sharon Horn Villasenor, and Mary Ann Renteria Mendoza; Gloria Fernandez, Matthias González, Adrian Davis, Oma Crockett, Mary Lou Pulido Garnica, Robert Horn, Petra González, Rachel Ramirez Calazada, and Gilbert and Betty Sosa.

We appreciate all the effort and work Image Source gave to help preserve some of these photographs for publication. Thank-yous also go to Jeff Ruetsche, Tim Sumerel, and Arcadia Publishing for their assistance and dedication to completing this project.

Unless otherwise noted, all images appear courtesy of the Sosa-Morales family collection. Images courtesy of the Fillmore Historical Museum are indicated by FHM.

INTRODUCTION

Gov. Jose Figueroa gave Rancho Sespe to Carlos Antonio Carrillo in 1833. This Mexican land grant consisted of 8,881 acres and included the Santa Clara River valley between Piru Creek on the east and Sespe Creek on the west; the mountains formed the boundaries to the north and south. It included present-day Fillmore. Carlos Antonio Carrillo was prominent in the Mexican political scene, and he was elected to the assembly. In 1837, he was appointed governor. Carlos was the eldest of seven children, and he and his brother, Jose Antonio Carrillo, soon were embroiled in California's revolutionary politics. After his defeat by Juan Alvarado and his forces, Carlos went on to serve on the Santa Barbara town council in 1849. Jose Antonio later became a member of California's Constitutional Convention, which ushered in statehood.

The Carrillo family took possession of the grant in 1842, and as required, they built an adobe house. Their primary residence was in Santa Barbara, and occasionally they would ride down to Rancho Sespe and host traditional Mexican fiestas. The adobe was located near Hall and Telegraph Roads, and it was partially destroyed by fire in the 1850s. During the late 1880s, children attending Santa Clara School, across the ravine from the old adobe, used to go and play among the ruins. Carlos Antonio Carrillo died in 1852, and his wife, Josefa, died in 1853. Thomas Wallace More and his two brothers, Andrew and Henry, purchased the entire rancho in September 1854 through an administrator's sale at the Santa Barbara Probate Court. The More brothers believed they were purchasing six square leagues or 26,000 acres; the US Land Commissioners had confirmed the ranch's size in April 1854: the Sespe grant consisted of six square leagues. But the US government appealed the approval based on evidence in a different version of the *expediente* that represented Rancho Sespe as only two square leagues. The Mores' attorney agreed to the government's two square leagues without the Mores' approval. The Mores raised sheep and cattle on the ranch. The ranch was surveyed in 1868 as two ranchos by surveyor Charles F. Hoffman, and the map was drawn in 1871. Thomas More received a title to his land in March 1872 for two square leagues, or 8,881 acres. The Craven Survey of Public Lands was filed on December 19, 1874, providing settlers with 90 days to file declaratory statements for lands on which they had settled, and this in turn opened the door for new settlers.

The California Agriculture Census indicates that by 1860, Thomas More had become the largest single landowner in Santa Barbara County, which at the time included all of present-day Ventura County. During this time, many settlers looked to claim land in accordance with the Homestead Act of 1862, and many of those people ended up near the junction of Sespe Creek and the Santa Clara River. Thomas More's ranch house was nearby, and soon a dispute arose with the settlers about the ownership of the land. More filed an application in 1875 to purchase the remaining four leagues of land from the government, but portions of this same land were already being claimed by homesteaders. More's request to purchase the additional land was denied by the Los Angeles Land Office, and as a result, the Sespe Settlers League was formed by the homesteaders to protect their claims. More claimed all six leagues for which he paid the Carrillos, but the settlers argued that the deed was fraudulent and that More paid for only two leagues. It was contested in the courts, and because the court system was slow and ineffective at the time, More proceeded to build a canal that diverted scarce water from the Santa Clara River and Sespe Creek onto his ranchlands before his petition to build the canal was granted. The More brothers dissolved their partnership, and Thomas More acquired Rancho Sespe and all the boundary controversies.

The drought years of 1876 and 1877 increased the tensions, and the settlers decided to take matters into their own hands. They knew they would be deprived of water for their own crops, so a group of them got together and came up with a plan. Under the cover of darkness, they met near the Thomas More ranch house on the night of March 23, 1877. Three of them entered the barn and set it on fire, while two of the masked men stood watch at the gate. The three then hid where they could see the house and the barn. More and his three workers rushed outside and evacuated the horses and tried to save some equipment. More returned to the burning structure and attempted to retrieve harnesses, but when he came out he was clearly visible and he was shot. He tried to get up, but several of the men hit him with bullets to his head. More lay mortally wounded, and the men escaped.

The crime scene was contaminated, since law enforcement footprints were mixed in with the killers' tracks, and the eyewitness accounts were later recanted. Seven men were charged with the crime, and two were convicted. The seven were J.S. Churchill, John Curlee, W.H. Hunt, Jesse Jones, J.D. Lord, Charles McCart, and Frank Sprague. Curlee was found guilty, although he was later granted a new trial and acquitted. Sprague was found guilty and sentenced to death. The sentence was commuted to life imprisonment, but he was pardoned and released after serving seven years in San Quentin Prison. Thomas More was disliked by the settlers in the area, and many did not see his death as a crime at all but thought he got what he deserved. While in prison, Sprague made cribbage boards out of wood, and the Fillmore Historical Museum has one of these on display in the Rancho Sespe Room. The Thomas More murder was considered the "crime of the century" at the time.

Before television came to Rancho Sespe, the children who grew up there played outside games such as kick-the-can and hide-and-seek, and they would go swimming in the Santa Clara River or Sespe Creek. Parents kept a watchful eye on their children while they played, and the parents sat on their porches talking and sharing coffee. At night, because there were no streetlights, the darkness enabled everyone to look up into the night sky and see the Milky Way and the various constellations. Many of the former residents recall when the Army sent German prisoners of war to Rancho Sespe to pick lemons and to other Ventura County locations to work in the citrus groves. At night, the prisoners of war were housed at the POW camp in Saticoy, California. Bobby Villanueva recalls that according to military records of June 1, 1945, the total POW population in the Saticoy Camp numbered 437. The camp was part of Cooke Army Base, where an additional 829 prisoners of war from Germany and Italy were based. In November 1956, Secretary of Defense Charles Wilson directed the transfer of 64,000 acres of North Camp Cooke to the Air Force; in October 1958, Cooke Air Force Base was renamed Vandenburg Air Force Base.

When television did arrive in the 1950s, good reception came from KNBC Channel 4 and KTLA Channel 5. In 1948, the Department of State brought several displaced Russian families to live in Rancho Sespe's Oak Village; they stayed for a short until they were moved to the Los Angeles area. Families shared Thanksgiving, Christmas, and other holidays together, and the children played outside. Some built forts and sometimes they would ride their go-karts on the Spaldings' driveway. This proved difficult when they could not control the speed and crashed into a neighboring barranca. Another common memory the various former residents have is when they would have to run away from the grazing Angus bulls kept in the pasture south of some of the homes. They tell how they had to beat the bulls across the pasture and would have to go under the barbed-wire fence that separated the pasture from the river. In time, the children grew up and families moved away. Rancho Sespe changed over the years, and after it was sold in 1979, the ranch's owners eventually eliminated family housing. A labor dispute arose between the management company known as Rivcom and the United Farm Workers labor union representing the ranch hands, but a contract was negotiated and the disputes were resolved in October 1987.

One

A Mexican Land Grant and the T. Wallace More Murder

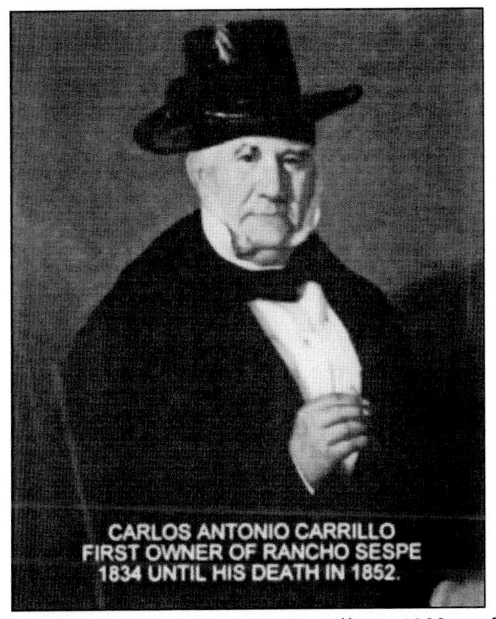

Rancho Sespe was granted to Don Carlos Antonio Carrillo in 1833, and he built an adobe south of today's Highway 126 near Hall Road. The adobe no longer exists. The sycamore tree at the intersection of Hall Road, Sycamore Road, and Highway 126 is California Historical Landmark No. 756. The bronze plaque located in front of the tree commemorates where John C. Frémont and his men passed on their march through the Santa Clara valley in January 1847. In 1825, Carlos Carrillo quit military service to enter politics. When the secularization of the missions began, Don Carlos was appointed *comissionado*, with orders to hand Mission San Buenaventura back to the Indians. Don Carlos Carrillo supported Gov. Juan Alvarado's revolution of 1836, but when Carrillo was appointed governor of California, Alvarado refused to leave his office. Alvarado, with a force of 200 men, engaged Carrillo's troops in a final showdown, and ultimately, Carrillo surrendered. Thereafter, word came from Mexico City designating Alvarado governor instead of Carrillo, and Carrillo retreated to San Diego. He died in 1852 at the age of 69, and his wife, Josefa Castro Carrillo, died one year later. Don Carlos Antonio Carrillo was buried in the Mission Santa Barbara cemetery. (Courtesy of FHM.)

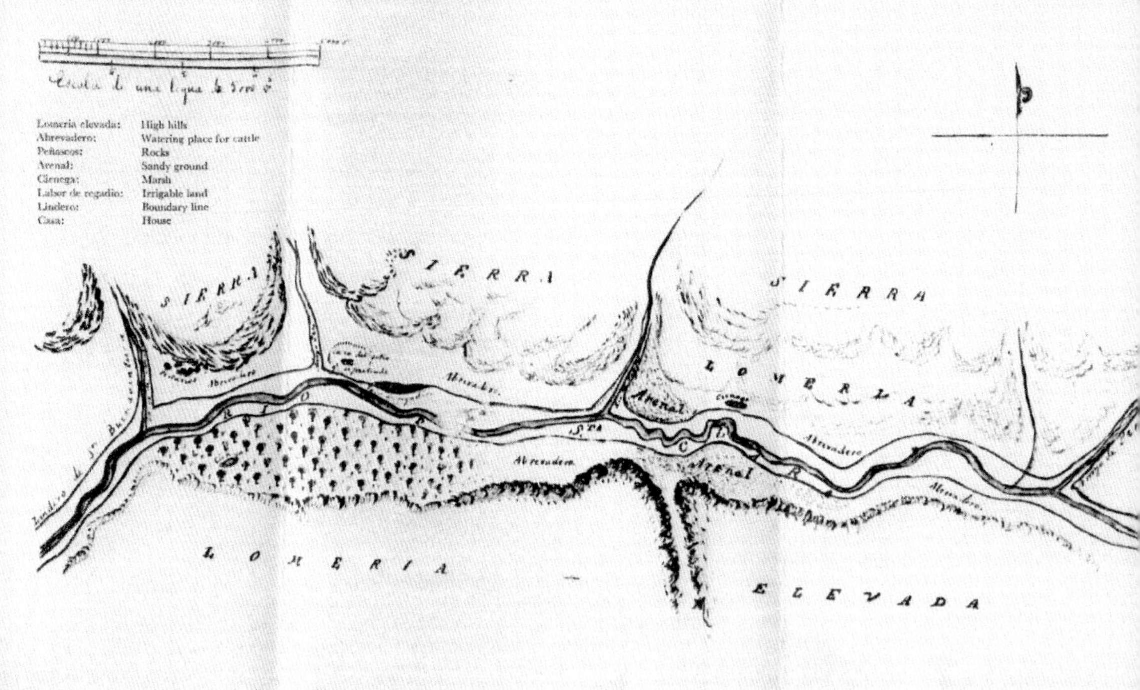

Thomas Wallace More and his two brothers, Andrew and Henry, purchased Rancho Sespe at the administrator's public sale in Santa Barbara. This is an early map of the Mexican land grant that Don Carlos Carrillo had made. The California Agriculture Census indicates that by 1860, Thomas More had become the largest single landowner in Santa Barbara County, which at the time included all of present-day Ventura County. During this time, many settlers looked to claim land in accordance with the Homestead Act of 1862, and many of those people ended up near the junction of Sespe Creek and the Santa Clara River. Thomas More's ranch house was nearby, ad a dispute soon arose with the settlers about the ownership of the land. (Courtesy of Huntington Library, San Marino, California.)

Thomas Wallace More was murdered late in the night of March 23, 1877. Seven men allegedly met at Frank Sprague's store at Rancho Sespe, used gunny sacks as masks, and made their way to the More house. Two of them positioned themselves at the gate of the corral while the others set the barn on fire. The commotion awakened More and his employees. More rushed out to the barn to save the animals. After he retrieved some harnesses, he came out into the open and was shot. As he tried to struggle to his feet, several of the men pumped shotgun lead into his head. More lay dead. Eventually, seven men were indicted. Sprague was the only man convicted, and he spent seven years in San Quentin before he was pardoned by the governor and released.

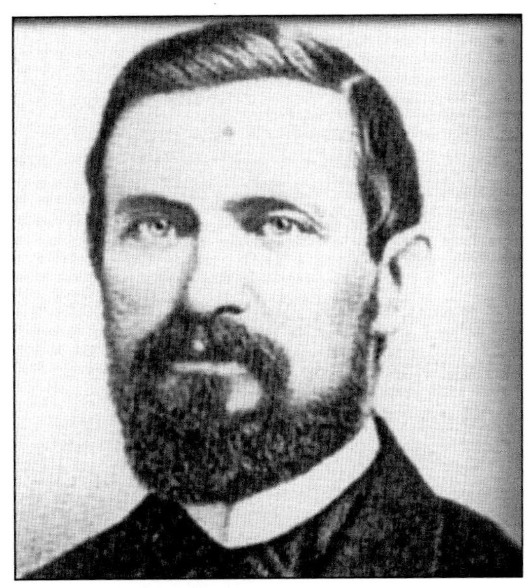

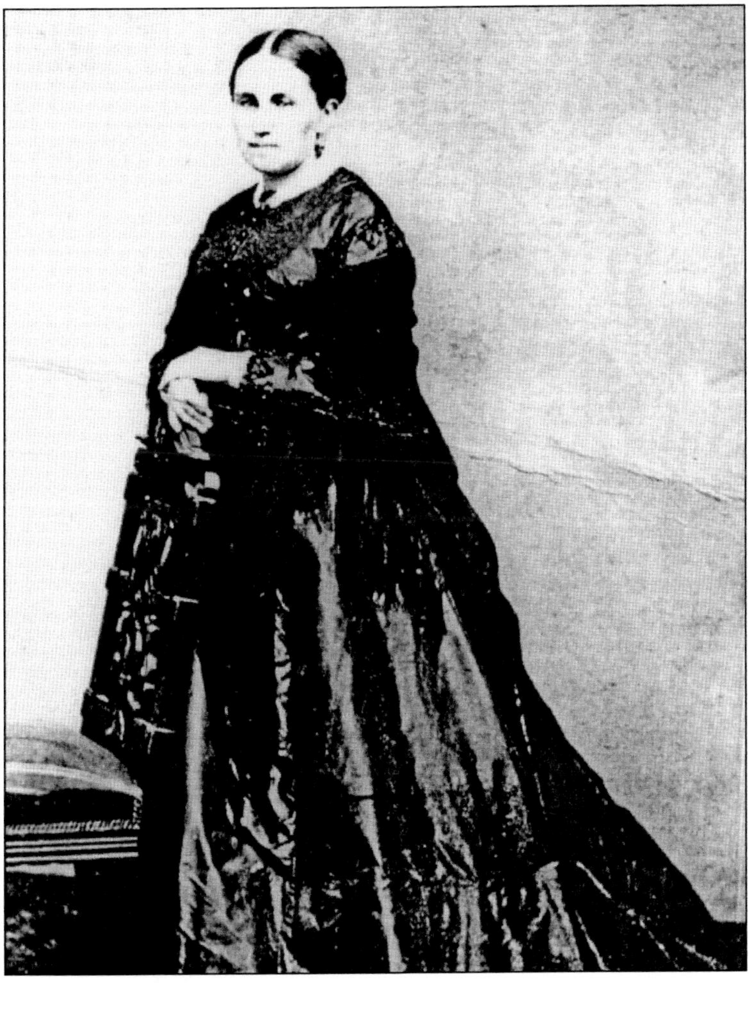

Susana Hill More, Thomas More's wife, was the daughter of Daniel Hill and Rafaela Sabina Luisa Ortega. Susana's maternal grandfather was Capt. Jose Francisco Ortega, who founded the San Francisco Presidio. He was also instrumental in founding Mission Santa Barbara. Susana and Thomas More had four sons and a daughter. Their daughter, Mattie, married Charles A. Storke, who assisted the More descendants with business transactions after Thomas's murder in 1877 and after Susana's death in 1879. (Courtesy of Evie Ybarra collection.)

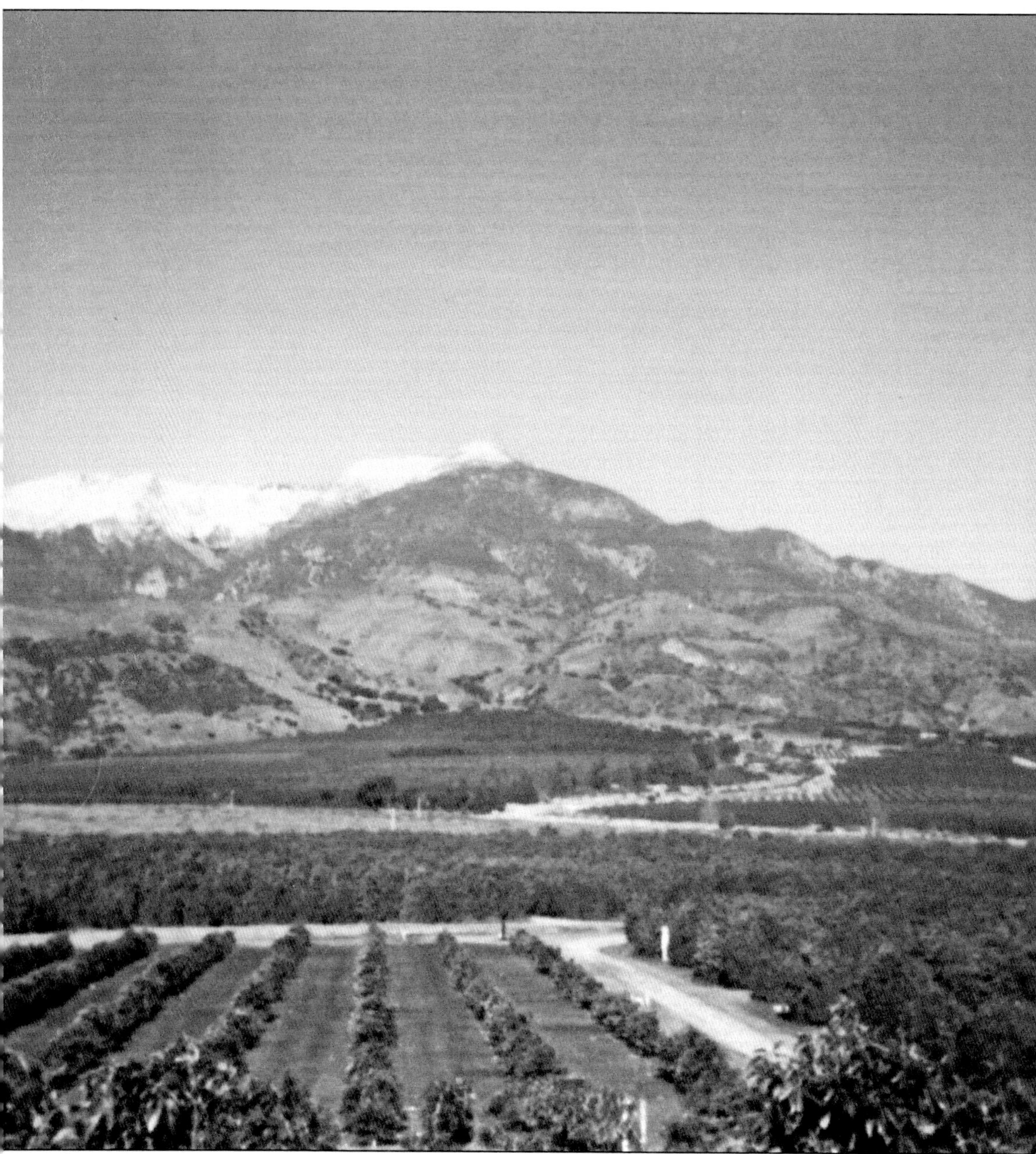

This photograph shows a view overlooking Rancho Sespe and facing north toward San Cayetano Mountain. California condors could be seen flying near San Cayetano Mountain before they became endangered. Now, this bird is protected and has an area labeled the Sespe Condor Sanctuary, which is back in the Sespe wilderness.

Two

Eudora Hull Spalding and Keith Spalding

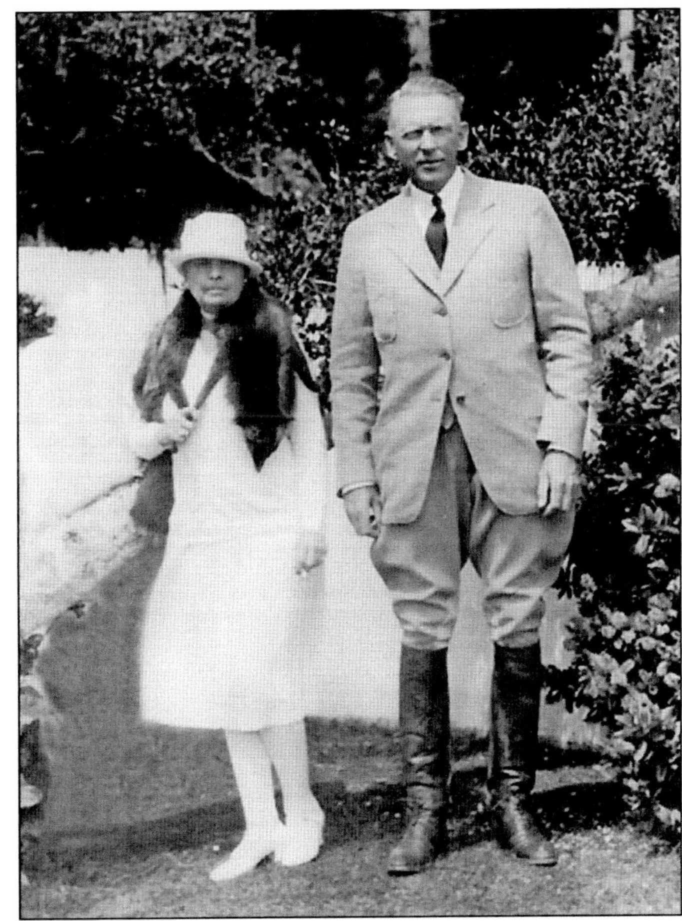

In 1888, Morton Hull from Chicago, a financier and businessman, foreclosed on the Rancho Sespe mortgage, and in 1895, he gave half of this property to his daughter, Eudora Hull, and the other half to his son, Morton. Later, Eudora purchased her brother's half, and she became the sole owner. Eudora and Morton were twins, born on January 13, 1867. Eudora married Keith Spalding on December 29, 1906; he was 10 years younger than she. Together, they made Rancho Sespe a working citrus ranch in the early 20th century. Keith and Eudora Spalding spent time living at Rancho Sespe and at their home in Pasadena. Married until her death in 1942, they enjoyed the outdoors and traveling to numerous locales throughout the world. (Courtesy of FHM.)

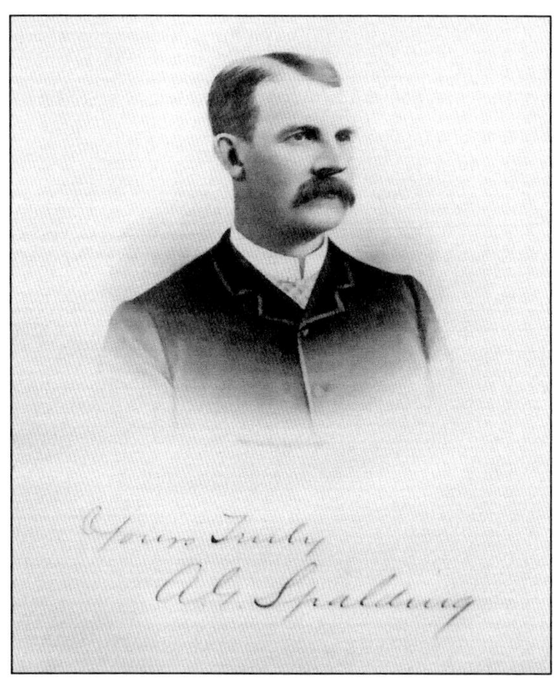

Albert Goodwill Spalding was Keith Spalding's father. Albert transformed the game of baseball into a professional sport. Albert was a star pitcher in his early years, and following the 1888 season, he led a group of 20 National League players on a "round the world" tour to eight countries, including Egypt and France. They posed with the Sphinx, and Albert viewed this trip as a public relations vacation, since he wanted good men whose clean habits would reflect well on the United States and the game of baseball. He designed sports equipment for the game and a rule book. He is the founder of the Spalding sporting goods company, which has since been sold. Albert Spalding died in 1915 in San Diego, and he was elected to the National Baseball Hall of Fame in 1939. (Courtesy of FHM.)

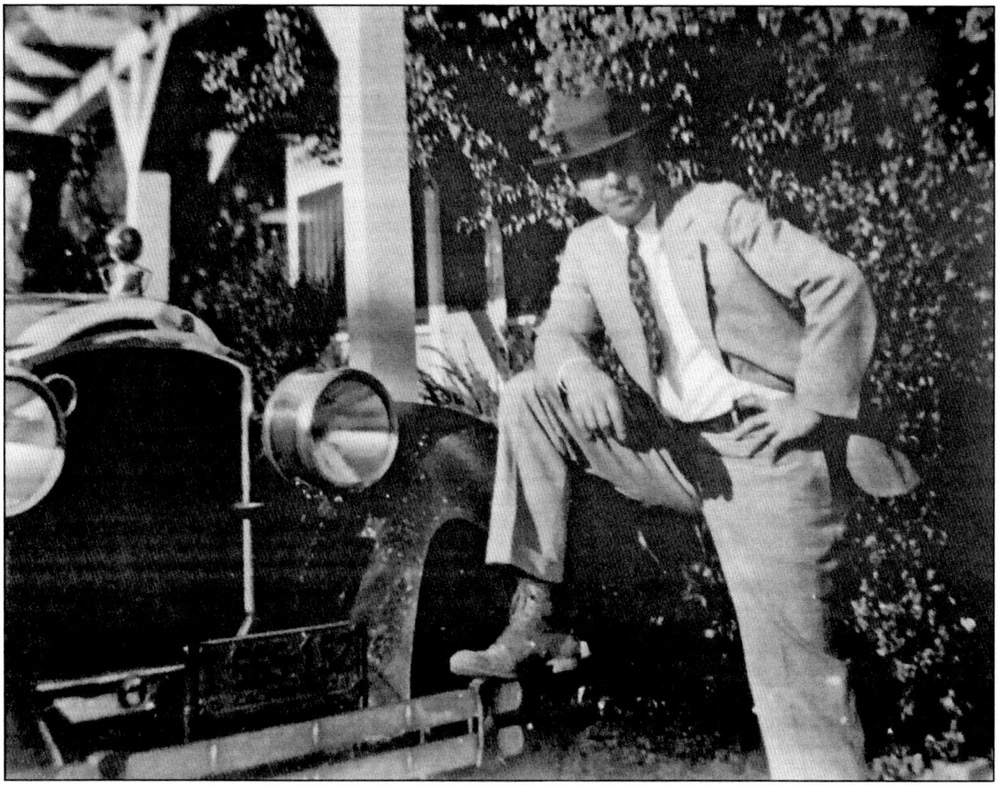

Keith is shown with his car in front of the Spalding bungalow on the Rancho Sespe property. He hired the architectural firm of Green and Green to design the bungalow and guesthouse as well as the first bunkhouse for the workers. (Courtesy of FHM.)

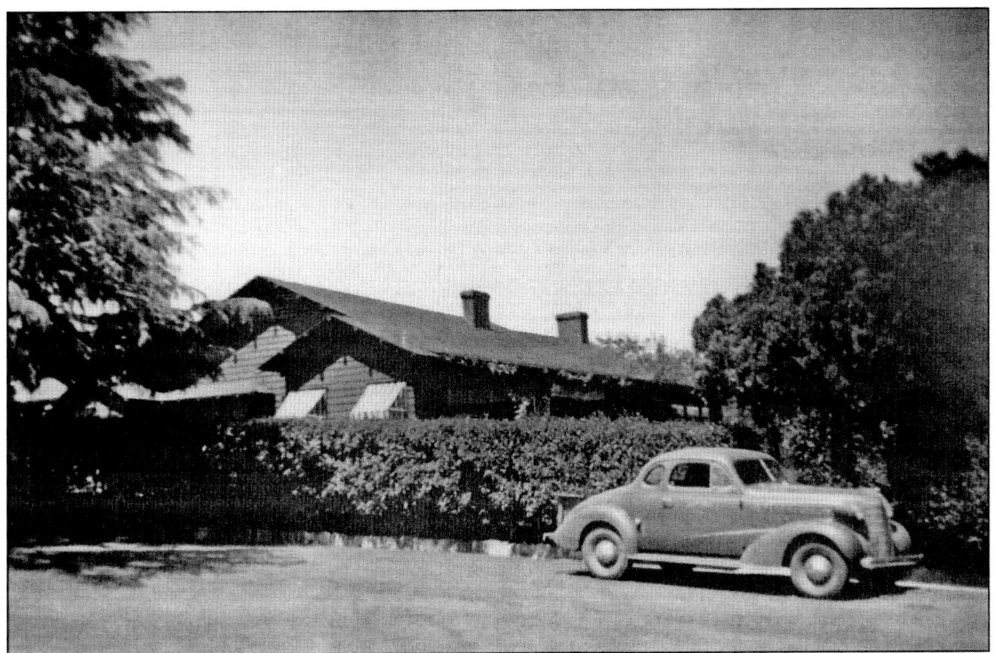

The Spalding ranch house was built in 1910 and it is now California Historical Landmark No. 135, along with the guesthouse and a stone wall on the Rancho Sespe property. Keith Spalding hired Harry Payton and Roy C. Wilson to build the ranch house. They constructed the single-story wooden main house and a nearby guesthouse. The main bungalow was a dark-stained wood until 1934, when it was painted white. Also at that time, a new chimney and an enlargement were added. (Courtesy of FHM.)

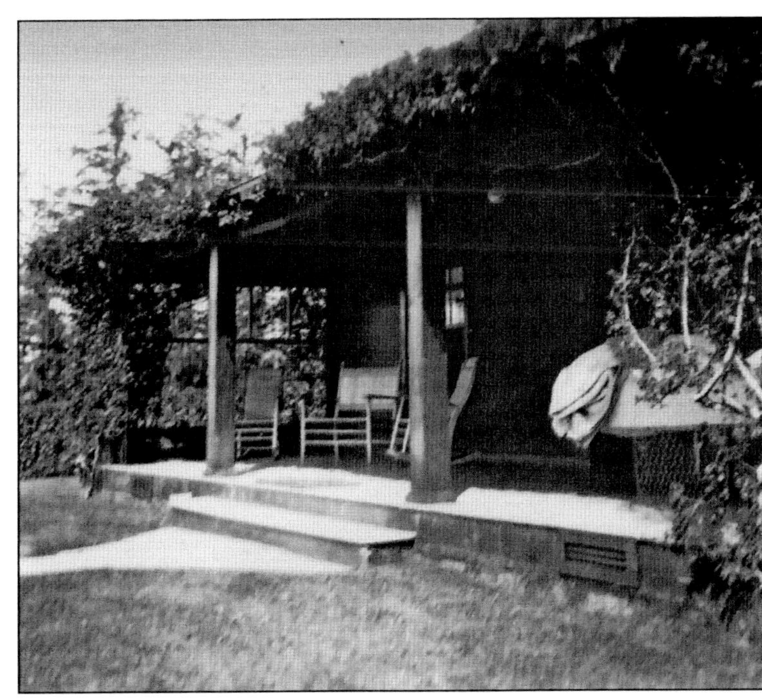

When the Spalding bungalow was painted in 1934, the homes in Sespe Village were also painted white. This was appreciated by all the families, since it brightened up the village. (Courtesy of FHM.)

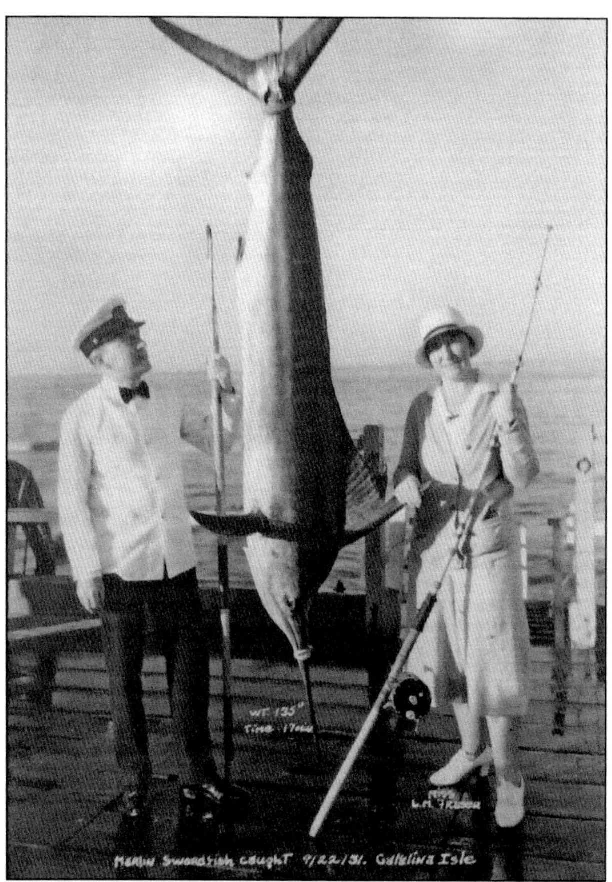

Eudora Spalding poses with a large tuna she landed. Eudora and Keith enjoyed fishing off the coast of Catalina Island and other places. One of the well-defined traditions is that as soon they caught a tuna or swordfish, they raised the appropriate flag. (Courtesy of FHM.)

This picture was taken with a view looking west on the eucalyptus-lined part of Highway 126. It is claimed that Eudora Spalding had the eucalyptus trees planted along the highway. She loved plants and trees of all types and enjoyed tending her garden and working in her greenhouse.

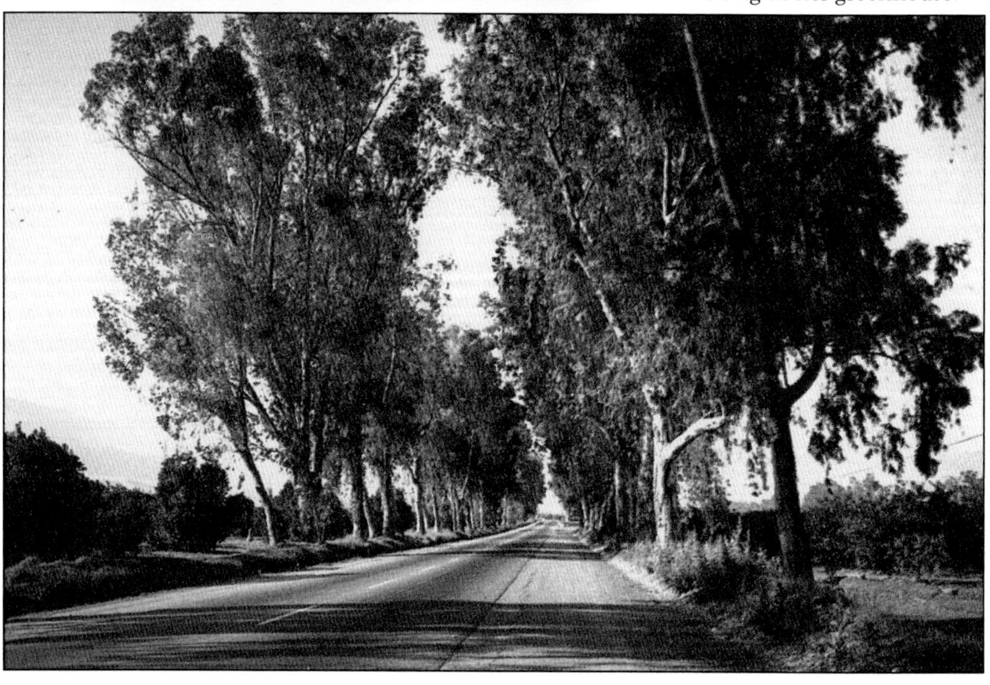

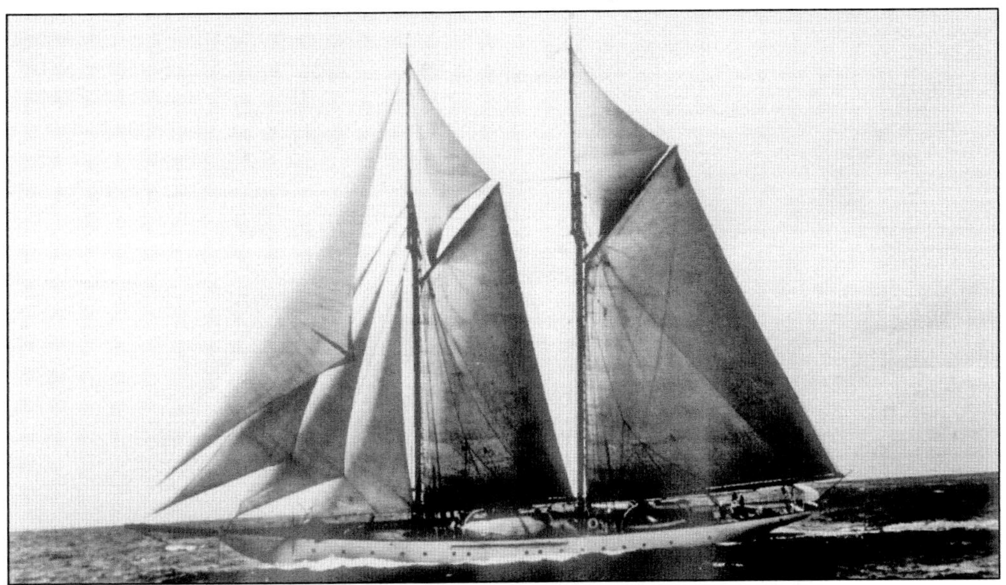

According to an article in the August 14, 2000, Daily Pilot, the Spalding schooner *Goodwill* is named after Keith's father, Albert Goodwill Spalding. This was one of the largest yachts ever kept in Newport Harbor in California. The 161-foot schooner was designed by H.J. Gielow, built by Bethlehem Shipbuilding Corporation in Delaware, and launched in 1922. The *Goodwill* crossed the Atlantic twice, and the Spaldings took her on an extended cruise to the South Pacific. In October 1942, the US Navy leased the *Goodwill* from Keith Spalding for $1 per year. The Navy commissioned her as a regular Navy vessel to operate on offshore patrol. A Spalding employee, Seaman Dahl, stayed with the schooner to be sure she was well-cared for, but when the Navy released her, all the teak wood had been painted gray, and the main cabin was riddled with holes from the countless dart games Navy seamen played in the formerly perfect cabin. The *Goodwill* made many trips to Mexican waters and to Cabo San Lucas. On her last voyage, she ran aground on a return trip to Ensenada from Cabo San Lucas. Tragically, nine people, including the owner, Ralph Larrabee, drowned.

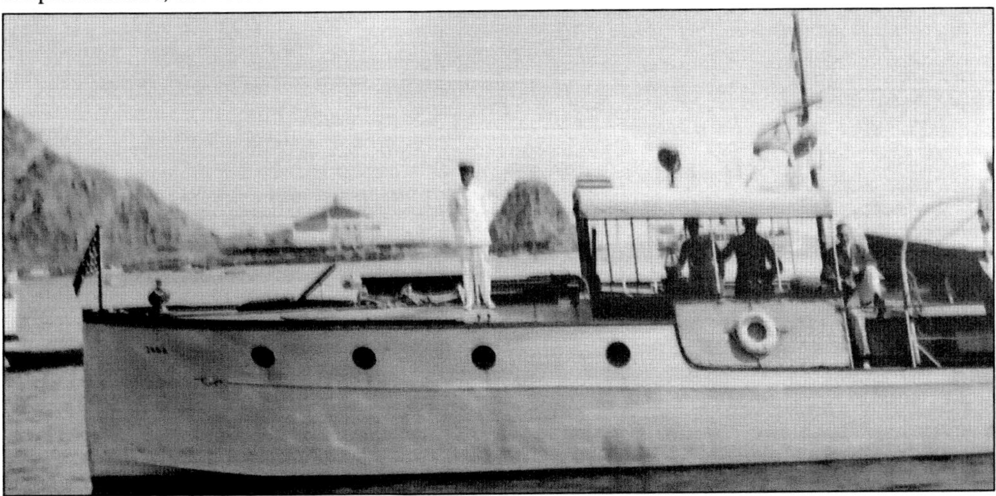

This is another of Keith Spalding's sailing vessels, a 54-foot Great Lakes Express cruiser equipped with a pair of six-cylinder Van Blerck motors. On one of their excursions, Eudora and Keith landed three tunas, all of them over 100 pounds. (Courtesy of FHM.)

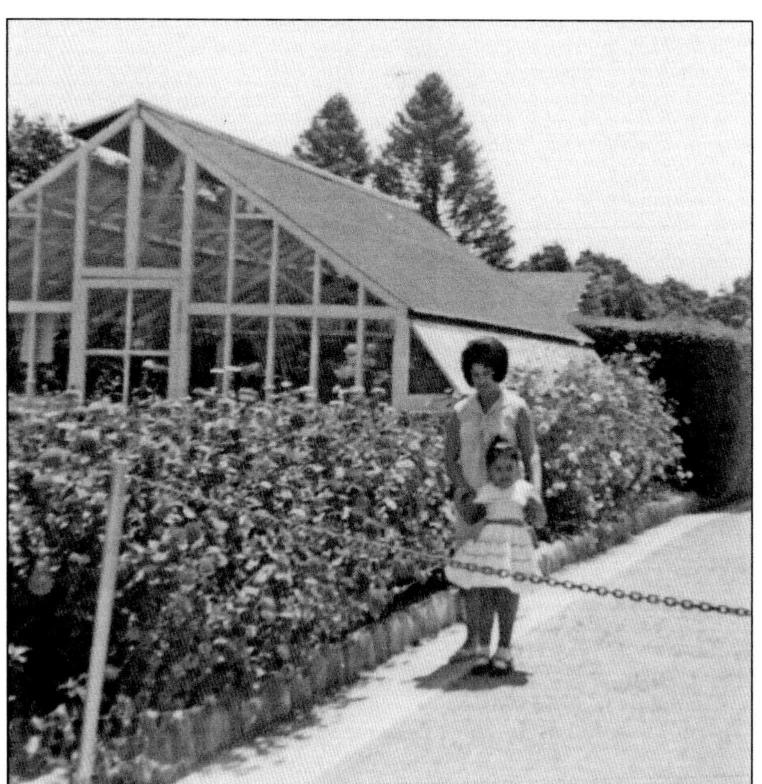

Carmen Torres and her daughter Norma are pictured in front of Eudora Spalding's greenhouse. Eudora enjoyed growing trees, flowers, and other plants. Carmen and her husband, Tomás Torres, lived in the guesthouse on the Spalding property and assisted the Spaldings with the management of the home and gardens on the property. (Courtesy of Tomás Torres family collection.)

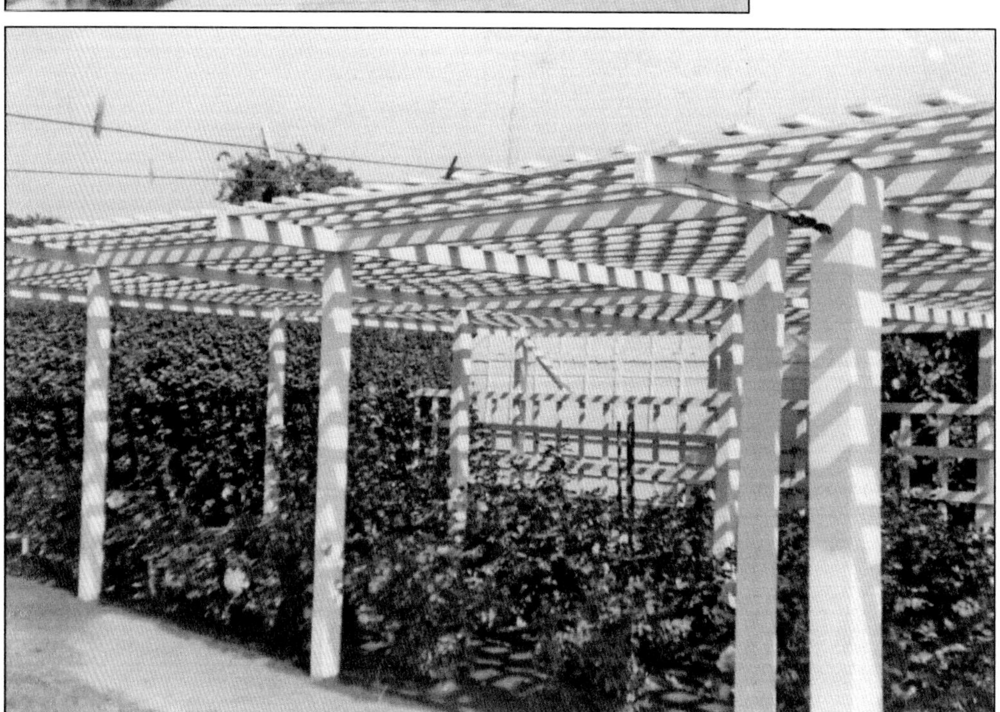

This is a photograph of Eudora Spalding's rose garden. Federico González and his wife, Felicitias, also worked at the Spalding property. (Courtesy of Larry Tepezano family collection.)

Tomás and Carmen Torres are standing in front of Keith Spalding's bungalow. (Courtesy of Tomás Torres family collection.)

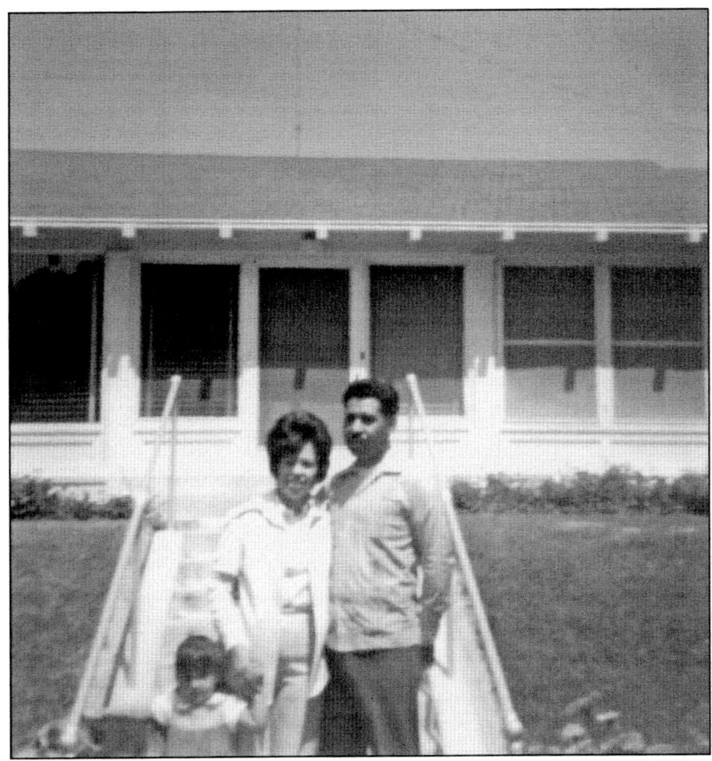

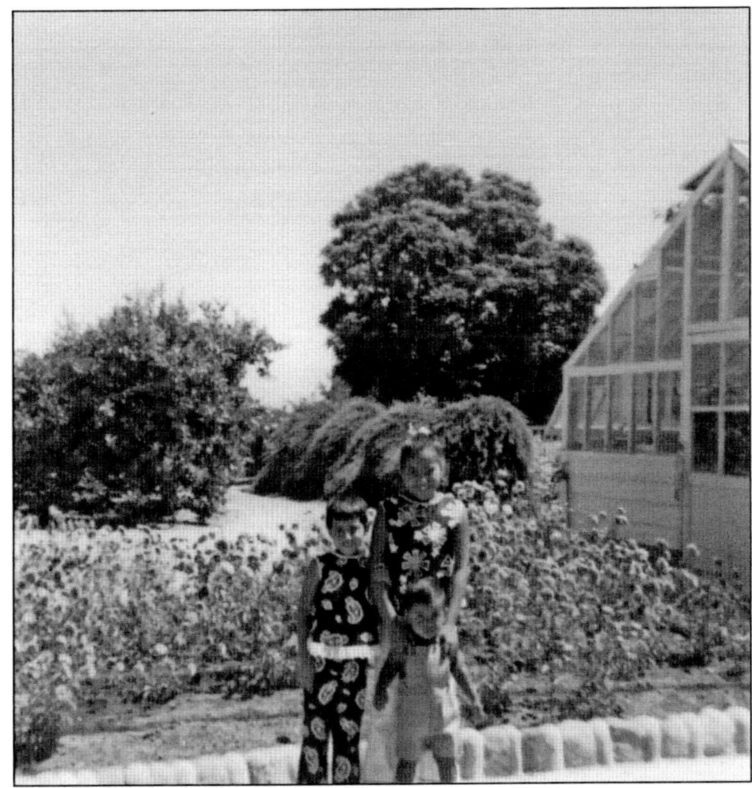

Norma and Cindy Torres used to play on the property next to their home, and one of their favorite places was near the rose garden. Eudora Spalding enjoyed her roses and the many plants in her greenhouse. (Courtesy of Tomás Torres family collection.)

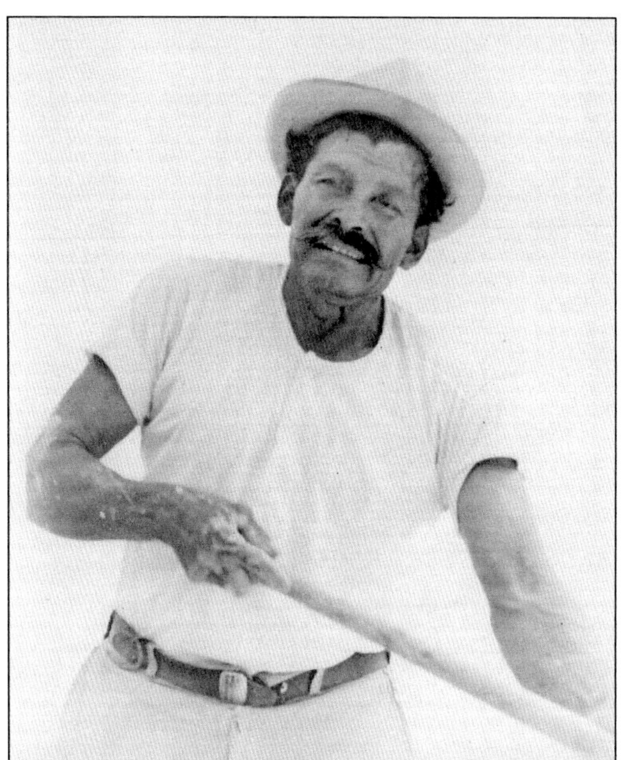

Alejo Davalos (pictured) and two others spent five years building the stone wall that lines the driveway up the hill to the Spalding bungalow. They quarried the stone slabs outside of Santa Paula and brought them to Rancho Sespe. Keith Spalding commissioned this 650-foot-long wall leading up to the main house in the 1930s. (Courtesy of FHM.)

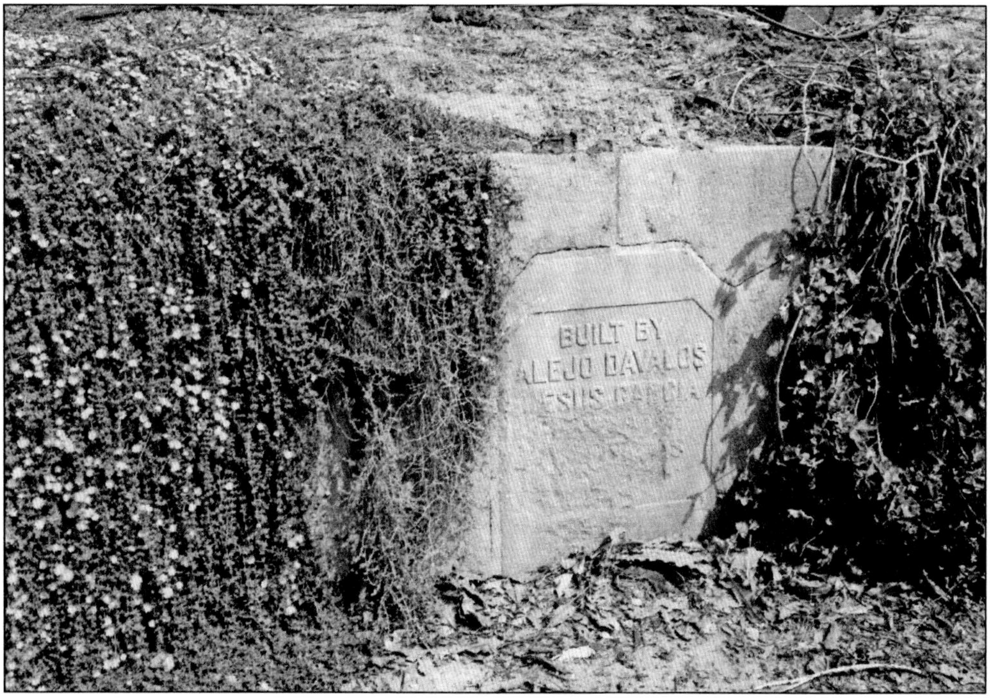

The Sespe wall is a historic landmark. The names of the three men who built it were engraved in cement along the stone wall, but the third name is obliterated. The names of Alejo Davalos and Jesus Garcia are still easily read. (Courtesy of FHM.)

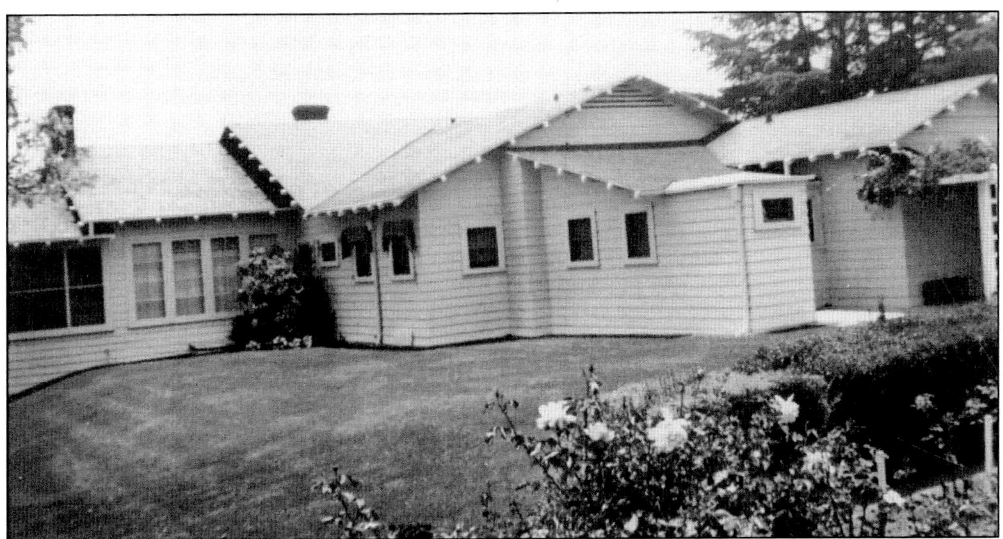

Here is an outdoor view of the Spalding bungalow. It is now a historic landmark. (Courtesy of Larry Tepezano family collection.)

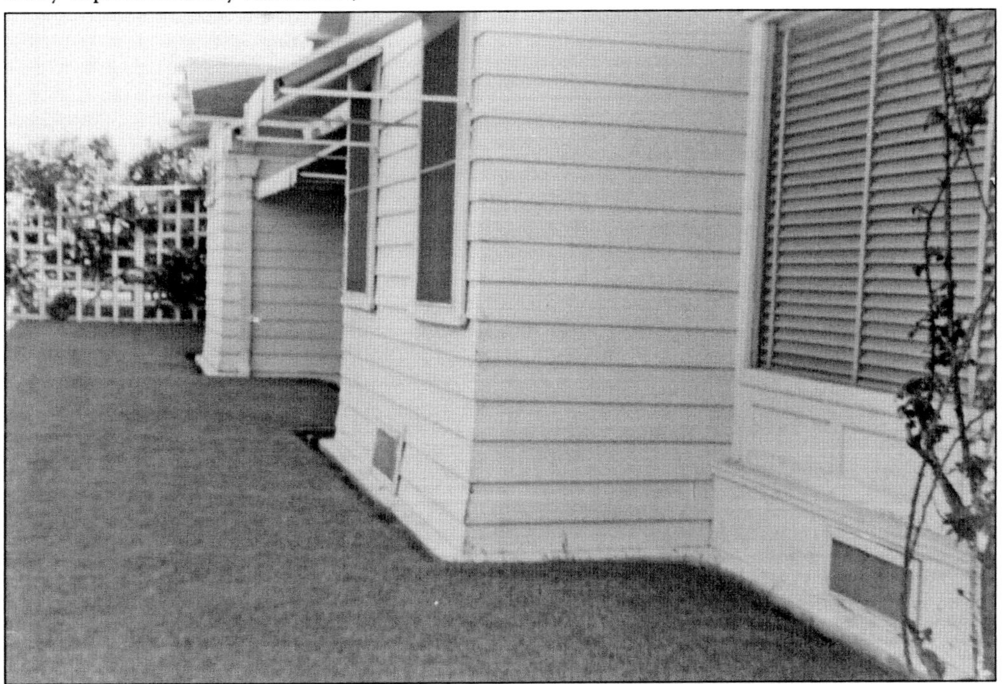

The first oranges, lemons, and walnuts were planted in 1908. The citrus industry is highly capital-intensive, and it attracted well-established farmers and businessmen from other parts of the country. Soon, these men established cooperative associations to handle all aspects of growing, packing, shipping, and marketing the fruit. The reputation of the citrus grower from Southern California was of a refined agriculturist whose hands need not touch the soil. During the early 20th century, Chinese, Japanese, and Mexican immigrants made up most of the labor force. Many Dust Bowl refugees arrived in the 1930s. The women who arrived worked in the packinghouses, and this was especially important after the labor turmoil of 1941 and after the relocation of the Japanese American population in 1942. (Courtesy of Larry Tepezano family collection.)

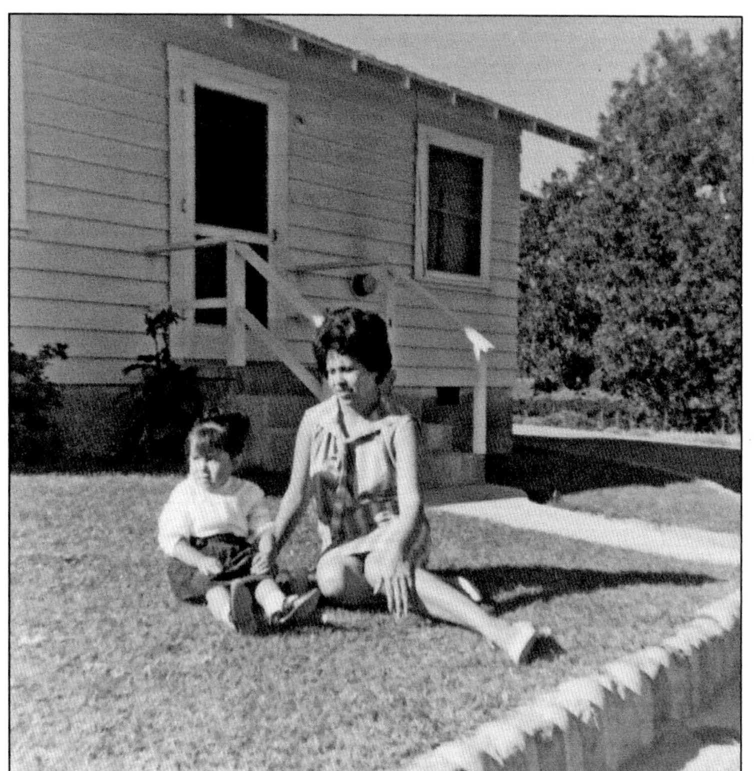

Carmen Torres and her daughter Norma are sitting in front of the bungalow they lived in on the Spalding property. This guest bungalow is a historic landmark. (Courtesy of Tomás Torres family collection.)

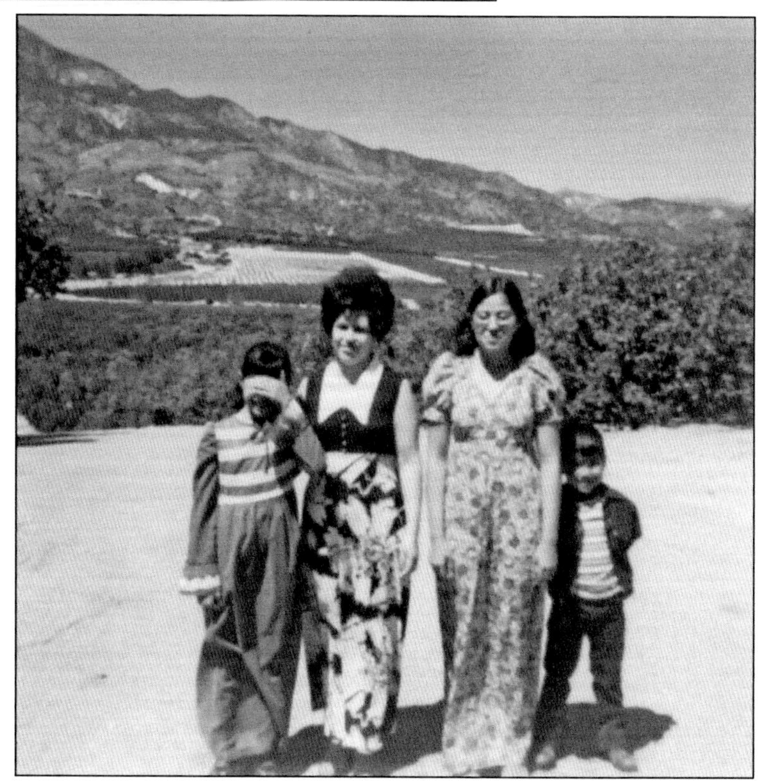

Rancho Sespe is shown from the Spalding guesthouse, where the children played. (Courtesy of Tomás Torres family collection.)

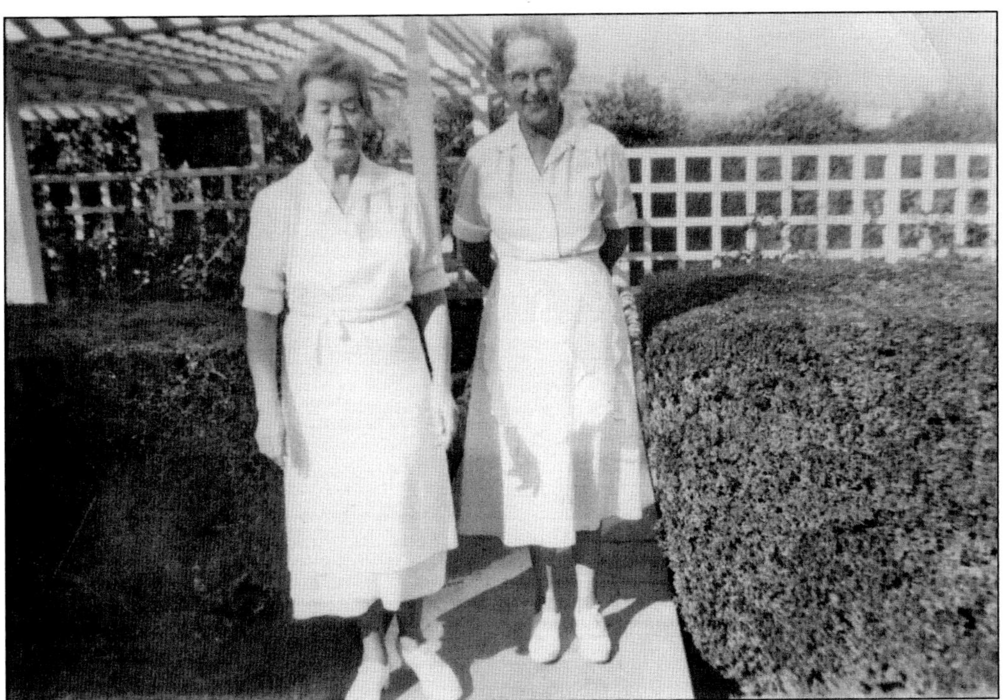

These two private cooks worked for the Spaldings. (Courtesy of Larry Tepezano family collection.)

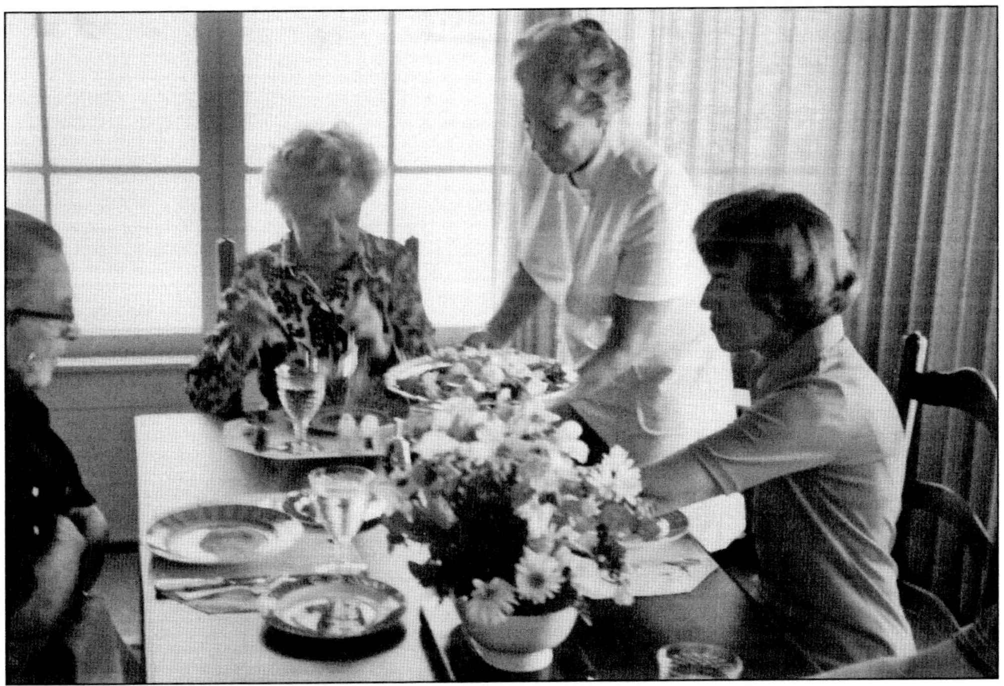

Connie Tepezano also worked as a cook for Eudora Spalding (second from left). Tepezano is shown serving dinner to Spalding and her guests. (Courtesy of Larry Tepezano family collection.)

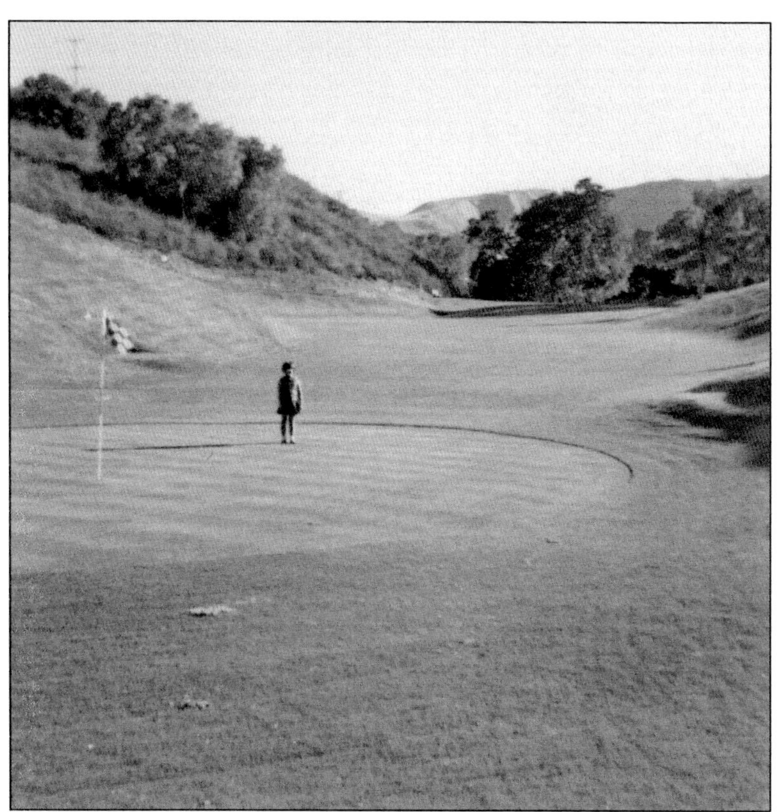
Norma Torres is shown on the Spalding golf course. (Courtesy of Tomás Torres family collection.)

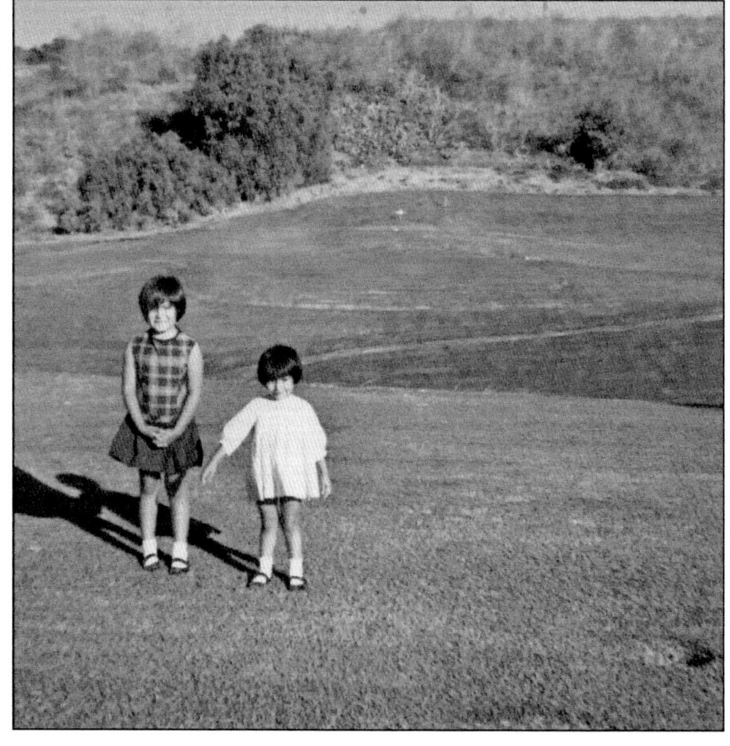
Norma (left) and Cindy Torres are featured on the Spalding golf course. In the 1970s, Woody Ybarra was invited to play golf there with some of the people who managed the ranch. The Ybarras attended a family barbecue at one of the homes then went up the hill to enjoy a game on the Spalding course. The golf course was just as it is in this photograph, very majestic and beautiful. One would never know of its existence, for it is hidden behind the hills just north of Highway 126. (Courtesy of Tomás Torres family collection.)

Norma Torres stands in front of the guest bungalow her family occupied. (Courtesy of Tomás Torres family collection.)

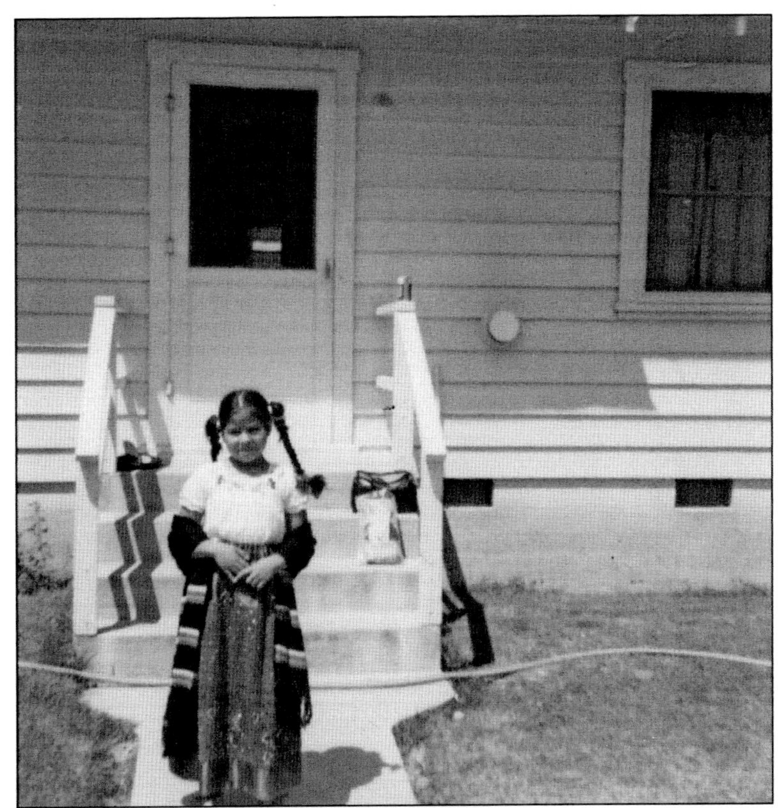

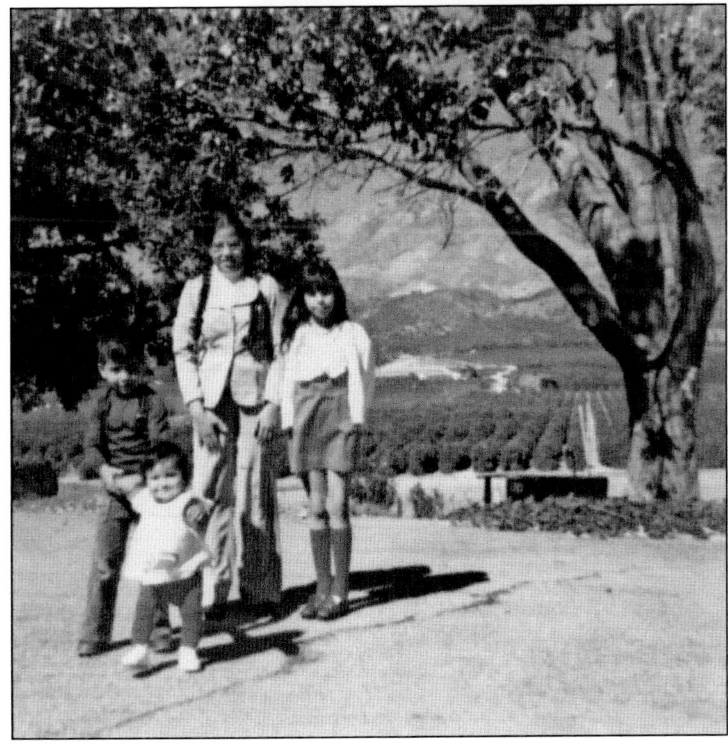

From left to right, Tomás Jr., Patty, Norma, and Cindy Torres pose for a photograph taken by their dad, Tomás Sr. (Courtesy of Tomás Torres family collection.)

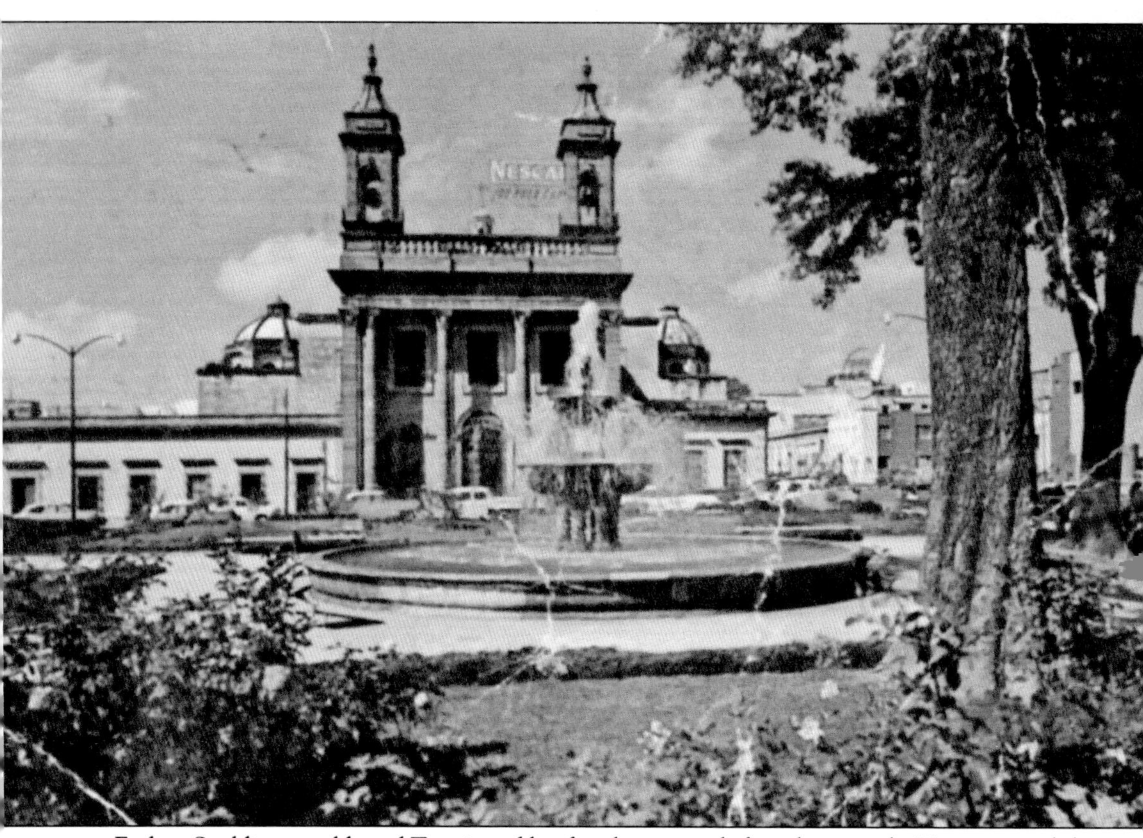

Eudora Spalding would send Tomás and his family postcards from her travels. Here is a card that was sent from Mexico. (Courtesy of Tomás Torres family collection.)

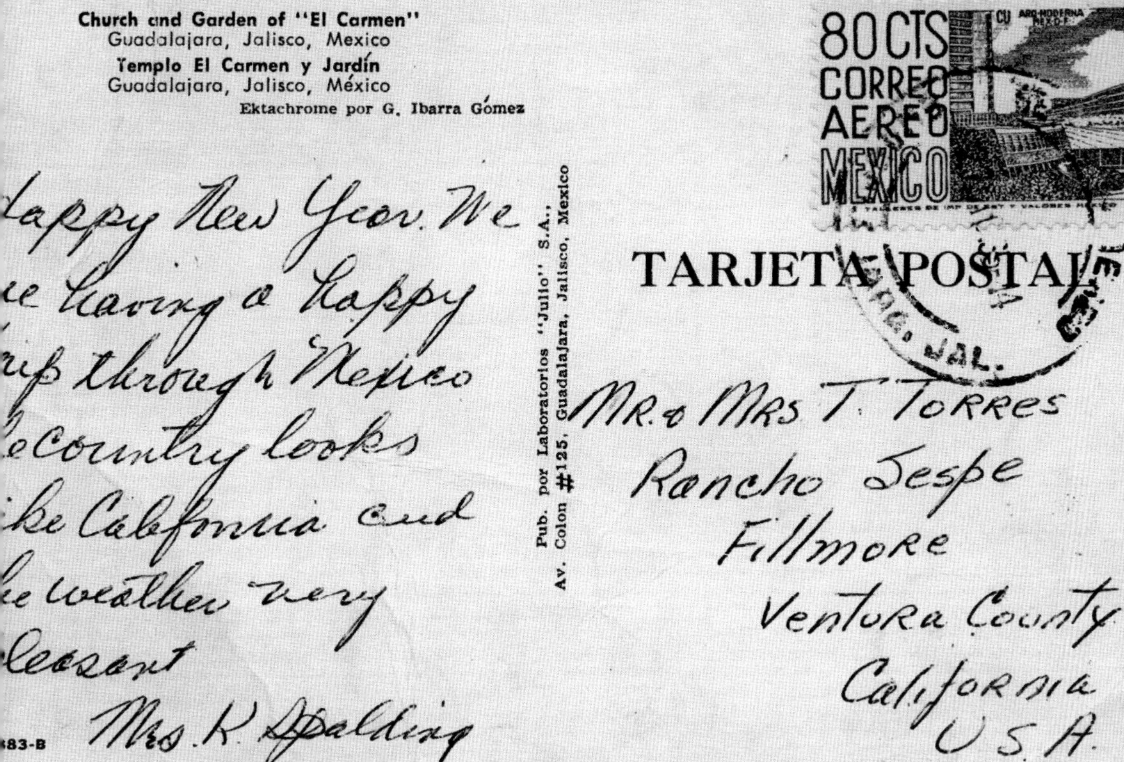

Church and Garden of "El Carmen"
Guadalajara, Jalisco, Mexico
Templo El Carmen y Jardín
Guadalajara, Jalisco, México
Ektachrome por G. Ibarra Gómez

appy New Year. We
e having a happy
rip through Mexico
e Country looks
ike California and
he weather very
leasant
Mrs. K. Spalding

83-B

Pub. por Laboratorios "Julio" S.A.,
Av. Colon #125, Guadalajara, Jalisco, Mexico

80 CTS
CORREO
AEREO
MEXICO

TARJETA POSTAL

Mr. & Mrs T. Torres
Rancho Sespe
Fillmore
Ventura County
California
U.S.A.

The Spaldings were traveling in Mexico and sent the postcard from Guadalajara with an image of the "Templo El Carmen y Jardín" (Church and Garden of El Carmen). Eudora Spalding's message reads: "Happy New Year, we are having a happy trip through Mexico. The country looks like California and the weather very pleasant. Mrs. K. Spalding "(Courtesy of Tomás Torres family collection.)

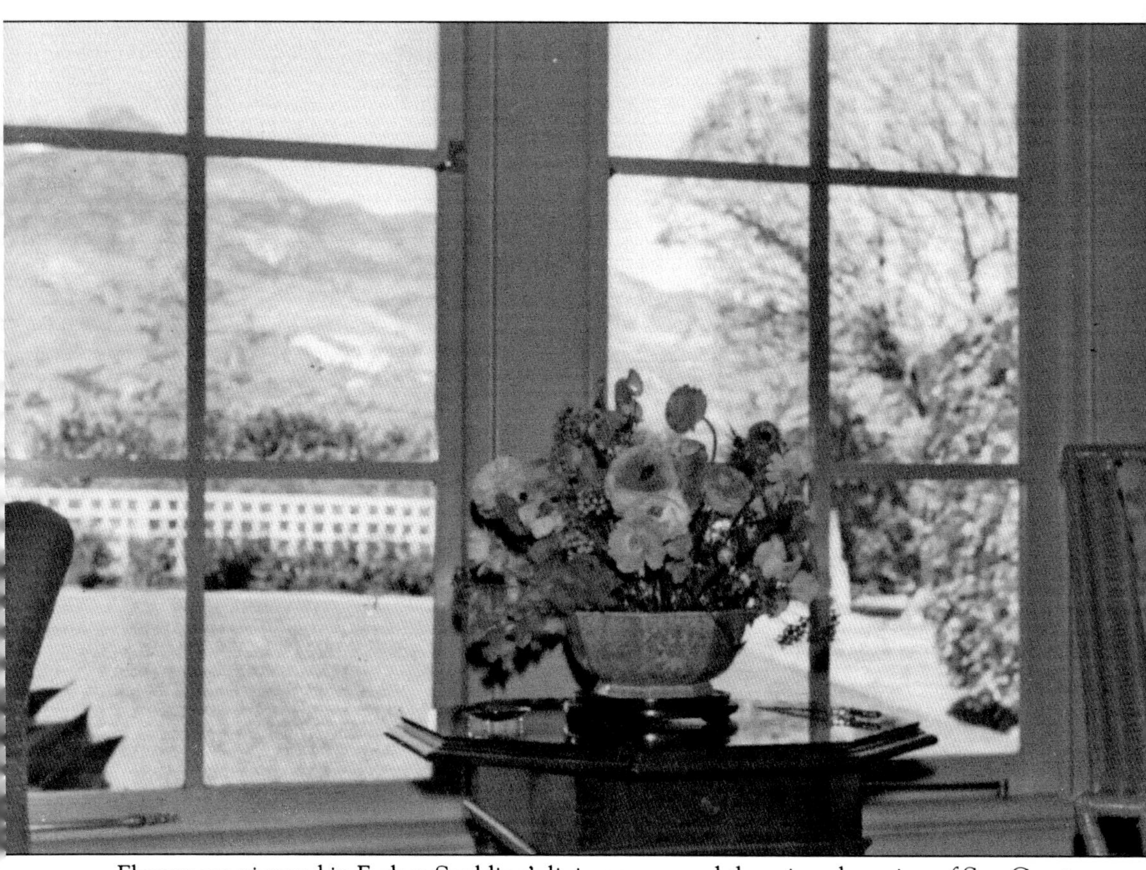

Flowers are pictured in Eudora Spalding's living room, and there is a clear view of San Cayetano Mountain. (Courtesy of Larry Tepezano family collection.)

Three

LIFE ON RANCHO SESPE

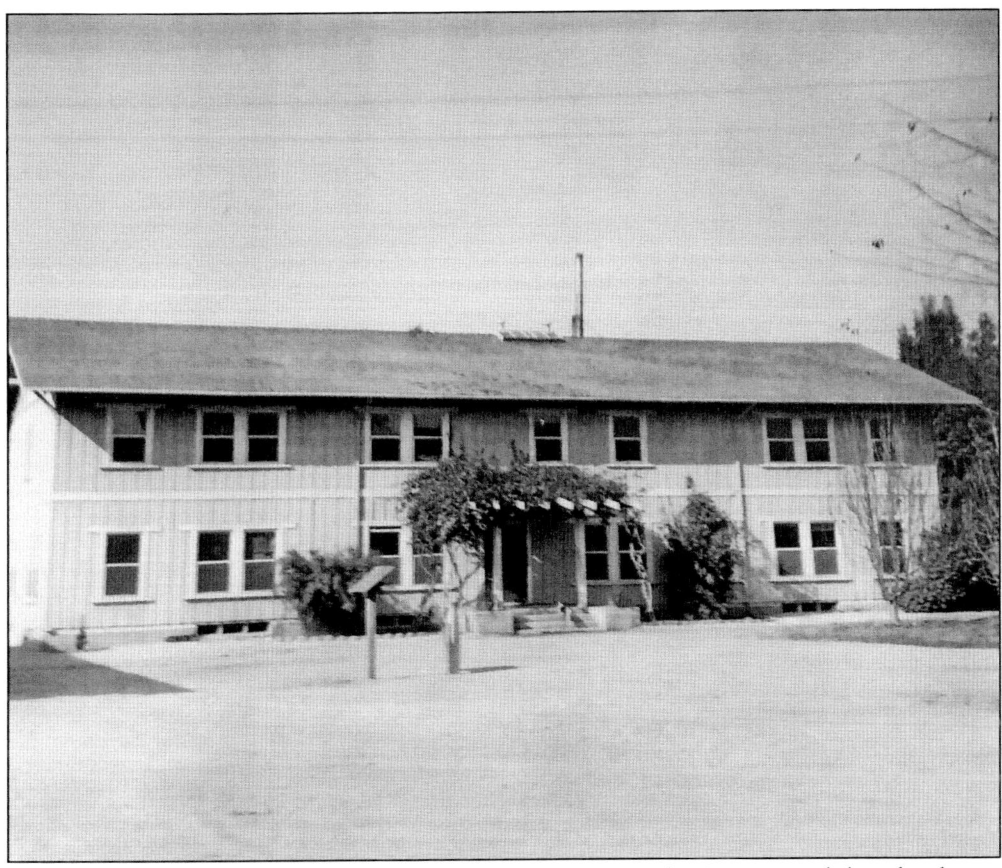

Here is the bunkhouse on Rancho Sespe. The laborers in Rancho Sespe reared their families in the workers' camp, which was known as Sespe Village (also called the Mexican Village), where 140 families lived. The Mexican workers were furnished with water and enough land for each house to have a vegetable garden. The homes were brown when they were originally built, then they were all painted white in the 1930s. Some families recall that they only paid $8 per month in rent and $4 for electricity. Others claim they did not pay rent, only utilities. In the May 1920 issue of *California Citrograph*, an article refers to the rent for married men ranging from $5 to $8 per month. Later, the rents were standardized to $15 per month. The Mexican families were encouraged to build their own houses, and $10 per month was taken out of their paychecks until the cost of the lumber was paid for. The philosophy was that the workers' housing had long-term benefits by maintaining a reliable workforce year-round. Families were stable and they brought relatives to live and work on the ranch as well. (Courtesy of FHM.)

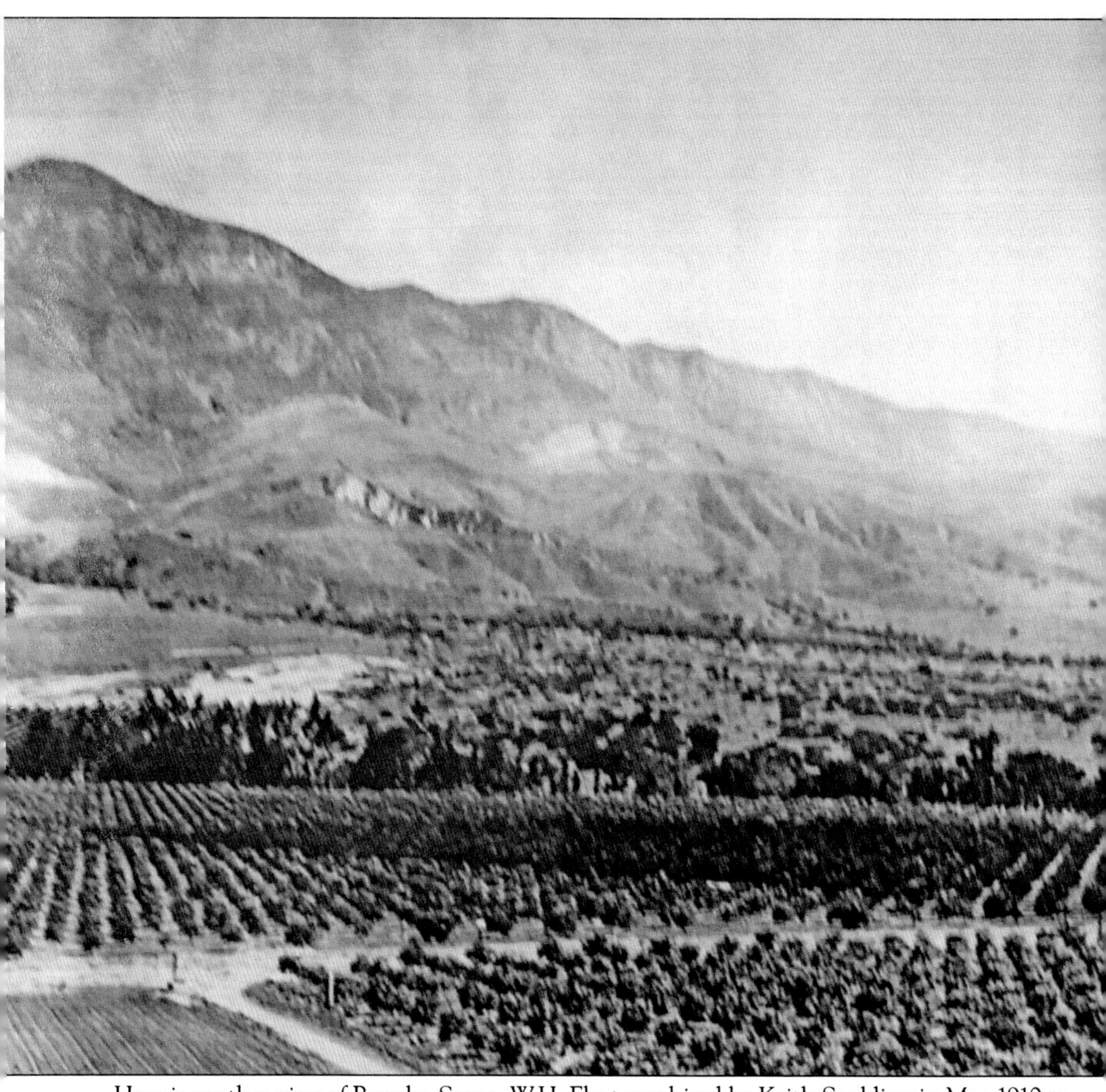

Here is another view of Rancho Sespe. W.H. Fleet was hired by Keith Spalding in May 1910 as the first manager of Rancho Sespe. He served through World War I. Fleet was responsible for the ranch's success from the beginning. He had the lemon packing plant built alongside the Southern Pacific Railroad, just north of Highway 126 and Telegraph Road. He was the overseer of the historic Rancho Sespe bunkhouses that were built for the Caucasian workers and designed by the well-known Pasadena architectural firm of Green and Green. He also started building homes for the operation managers and bookkeeper as well as for families. As acreage for the fruit orchards increased, more labor was required, and Fleet had the housing increased to accommodate all the workers and their families. According to Kenneth Glenn, retired assistant manager and executive vice president of Rancho Sespe from 1939 to 1978, the Spaldings intended to make Rancho Sespe into a showplace, and they "plowed the profits back into the property." (Courtesy of FHM.)

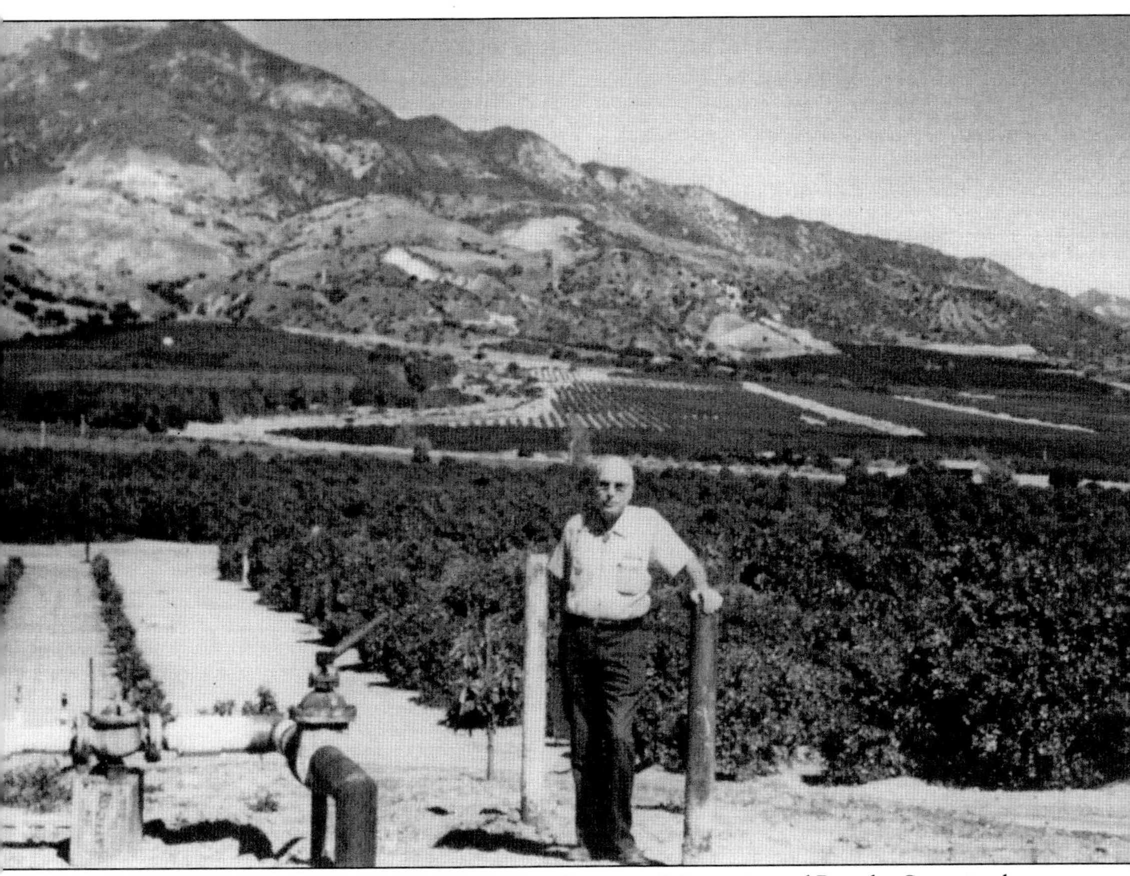

Kenneth Glenn is featured here with the vast San Cayetano Mountain and Rancho Sespe in the background. The *hacendado* and planter did not view the world in capitalist terms; instead they interpreted their world through the attitude of paternalism—they viewed themselves as attaining enough to sustain their lifestyle and to provide for their laborers. The landowners did not seek wealth for its own sake but instead valued their reputations as landowners and as "owners" of laborers (sometimes called peons) to work their large ranchos. Landowners considered their haciendas and labor force a mark of honor and prestige that confirmed their social status. Vincent Perez, in *Remembering the Hacienda History and Memory in the Mexican American Southwest*, discusses this cultural attitude: "Membership in the ruling class took precedence over the accumulation of capital." This attitude was embraced by the former owners of the Mexican land grant ranchos, but the new thinking was to make profits that could be sustained. (Courtesy of FHM.)

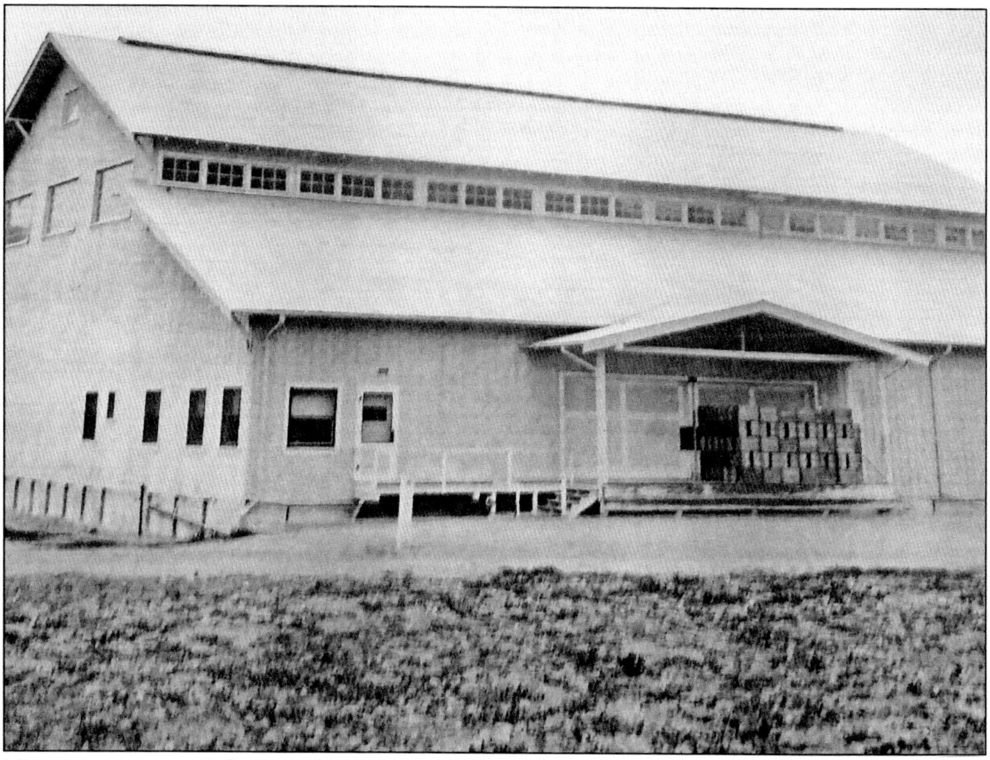

Here is an image of the Rancho Sespe packinghouse, which stood just north of Highway 126 and Telegraph Road alongside the Southern Pacific Railroad. This was built in 1916 under the direction of Rancho Sespe manager W.H. Fleet. Whenever Keith Spalding needed a train to stop, he would wave a wooden sign with his name painted on the front. (Courtesy of FHM.)

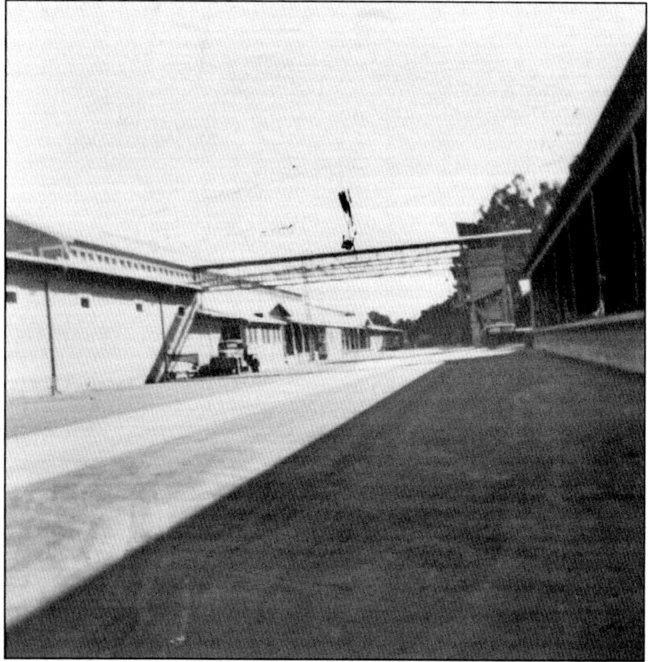

This is another view of the Rancho Sespe packinghouse. It handled over 4,000 railroad cars of fruit per year at the peak of production. The fruit was shipped all over the world. Olivia Carrera Lopez recalls the story about her mom, Esperanza Silva, who moved to Rancho Sespe in the 1920s. She had two siblings—Jesse, her brother, and Marie, her sister. Esperanza and her mother, Maura Silva, worked at the Rancho Sespe Packing House. Later, Esperanza had three children of her own—Joe Pedroza, Angel Carrera, and Olivia Carrera Lopez. In 2017, Olivia was elected as city clerk for the city of Fillmore.

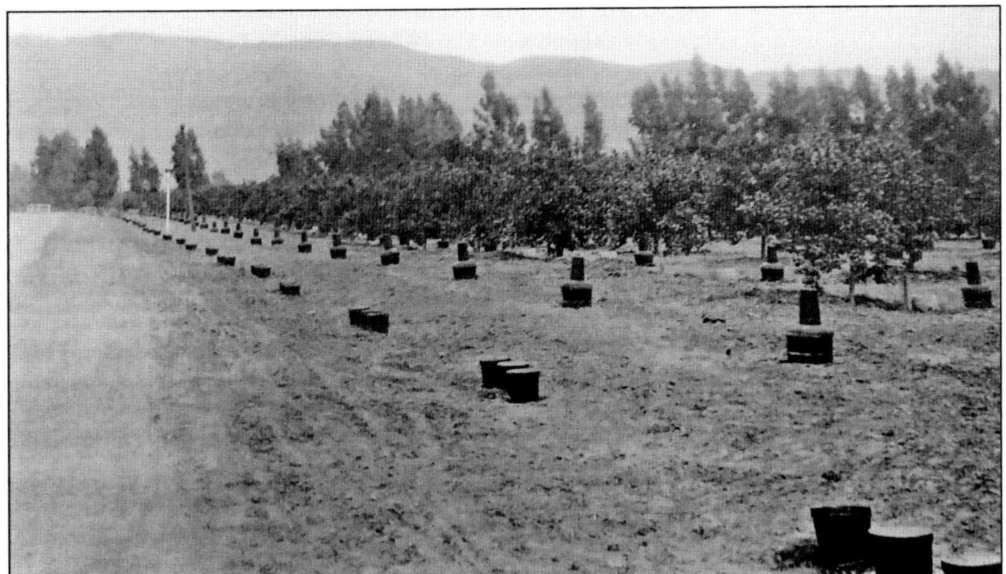

Smudge pots were used whenever the temperature dropped dangerously low. There was no frost protection on the ranch prior to the 1913 freeze, which killed many of the trees. Heaters were purchased after that freeze, and the first were of five-gallon capacity and very inefficient. After many years, according to Kenneth Glenn, the University of California, Riverside, developed the nine-gallon return-stack heater, which burned fuel without smoke. Forty heaters were placed per acre in orange and grapefruit orchards and 50 per acre in lemon and avocado plantings. In the 1960s, wind machines were placed in some of the colder orchards to save on oil. Rancho Sespe had both electric wind machines and gasoline-powered machines. For many years, there were 180,000 heaters and 22 wind machines. (Courtesy of FHM.)

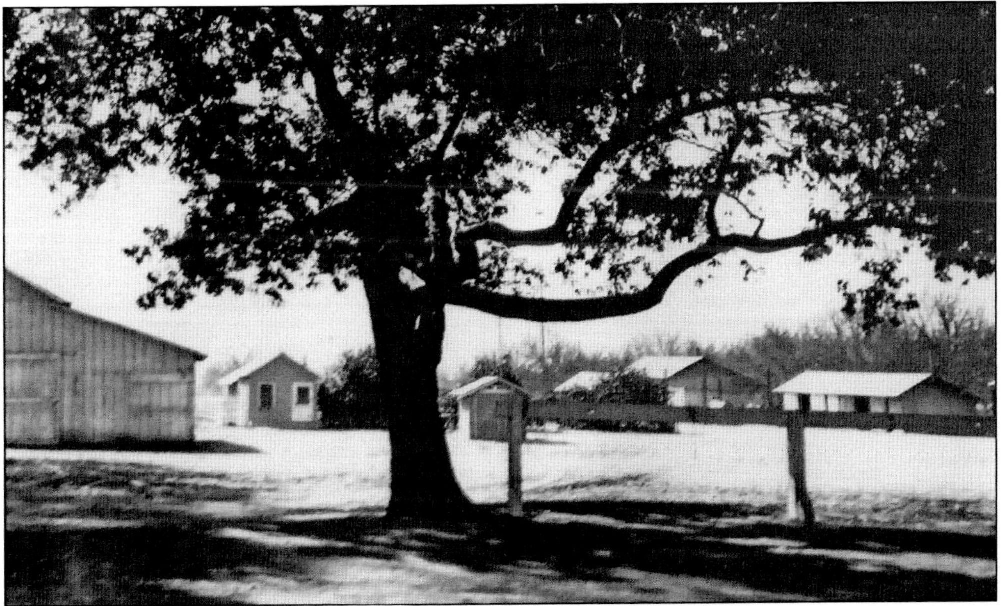

This is the complex of buildings on the ranch. During a general freeze, it would take 100 or more men to light up the heaters at night and the same number the next day to fill the heaters that had been burned. In the 1937 freeze, heater oil was purchased for 3¢ a gallon. (Courtesy of FHM.)

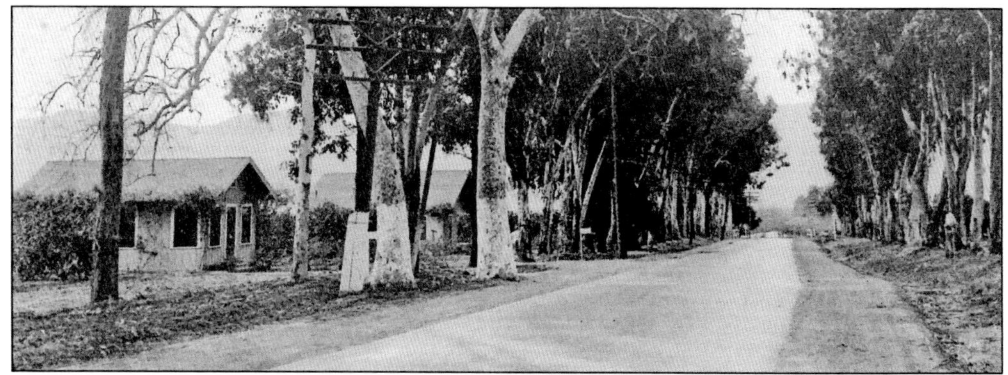

Housing for the bookkeeper and the Rancho Sespe manager is shown here. The Californios' attitude differed from the Americans' because many Americans determine social status by an individual's accumulation of wealth and property. The Californios, however, produced enough wealth for their lifestyle, which guaranteed their social position as a member of the ruling class of landowners. This promoted a natural dependency between the owners and the workers, because the workers were dependent on the owners' generosity, food, and shelter. The landowner depended on the workforce to tend the fields, the cattle, and the harvest and to bring food to the table. On Sundays, according to Francisco Perez, Keith Spalding and his ranch manager would ride their horses to inspect the village and to greet the workers. Spalding inspected his property regularly to find out the needs of the workers. (Courtesy of FHM.)

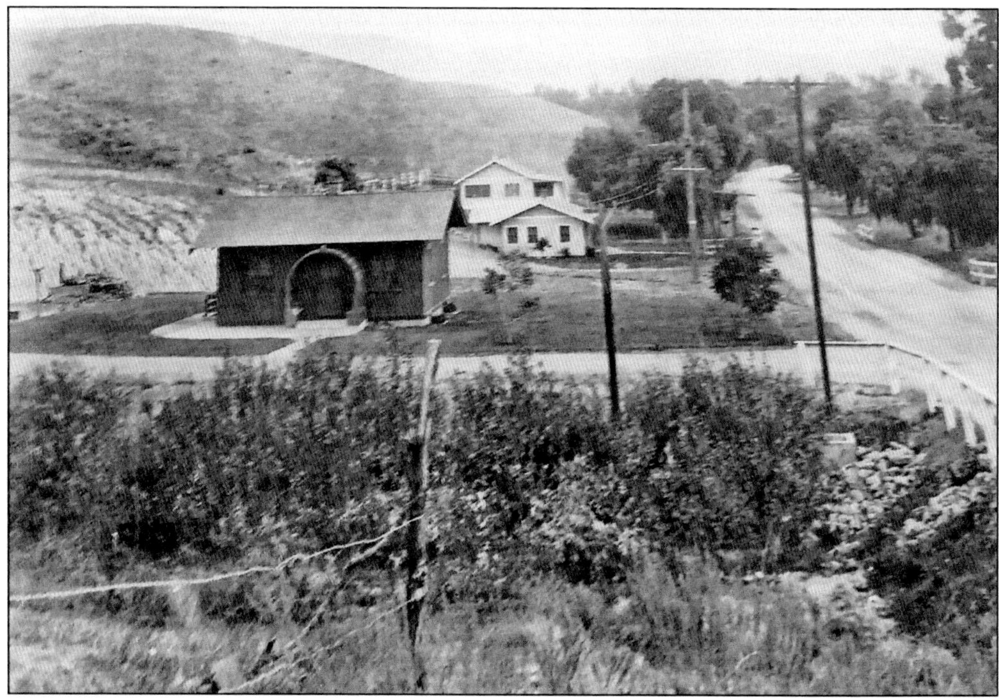

More housing was built during the 1920s and 1930s because of the increase in fruit production. Sespe Village had 140 single-family homes, a store, and a recreation building or dance hall, as many remember it. In 1946, Oak Village was built from surplus Navy barracks that were converted into two-bedroom duplexes. This increased the housing by 60 units, which then made a total of 200 housing units to handle up to 350 workers during the peak summer picking season. (Courtesy of FHM.)

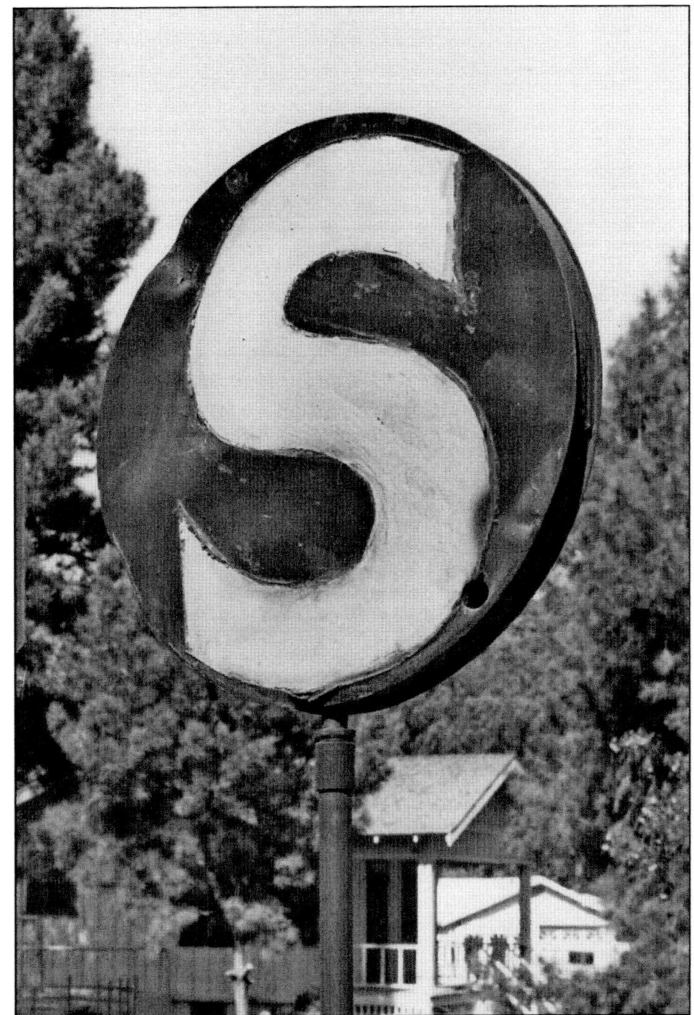

This *S* marked the entrance to the Spalding driveway. During this time, all the families living and working at Rancho Sespe were Spanish-speaking, and it was important to the families that their children learn English so they could find success in their new country. Petra González explained that a committee was formed and they decided to hire a teacher, Señora Artea, to instruct the children in Spanish, so they were bilingual. They were taught reading, spelling, writing, history, geography, folk songs, dancing and Mexican crafts. Eventually, the community of Rancho Sespe began celebrating Cinco de Mayo and September 16 on a large scale. (Courtesy of FHM.)

This is the sign that travelers saw upon entering the Rancho Sespe property. The Spaldings built an open-air dance hall with benches all around and a stage on one side. Programs and plays were produced there. The dance hall was in the center of the village, and all the parades started there. They would follow the main thoroughfare towards the entrance of the ranch and around a large tree, then back to the dance hall. Children would decorate their wagons, and a marching band or Latin musicians would play for the event. The special events were known all over the county, and people came from Oxnard, Ventura, Santa Paula, and Fillmore for a day of celebration and fun. (Courtesy of FHM.)

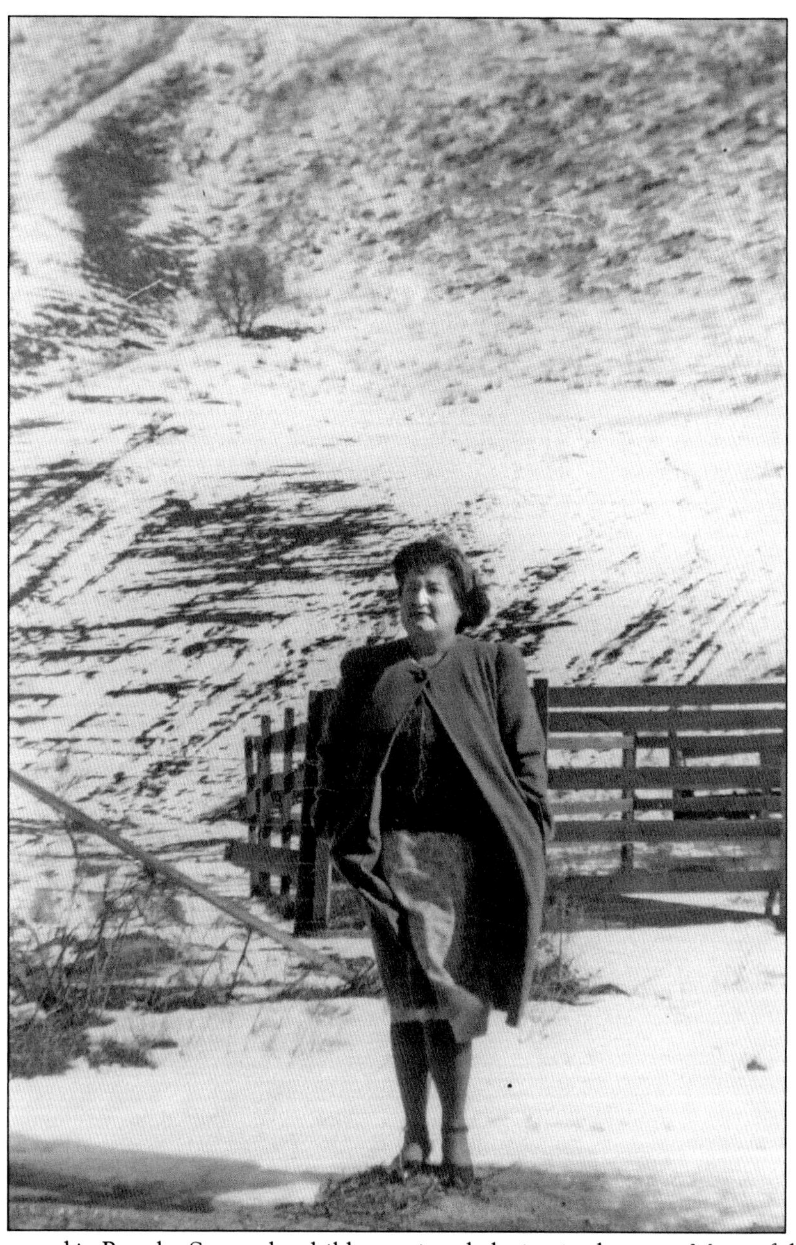

When it snowed in Rancho Sespe, the children enjoyed playing in the snow. Many of the women worked in the packinghouses while their husbands worked in the fields. Some shared the story of how Dorothy Peyton and her family would assist many families in Rancho Sespe. Dorothy drove to Sespe Village to distribute produce her family raised on their ranch. She offered her services as a taxi, taking anyone who needed to see the doctor or to go into the town of Fillmore. Dorothy also spoke Spanish and was an effective translator. The Peytons owned a walnut orchard, and Harry Peyton contracted with Petra González's family to pick the whole orchard. All the González children would pick the walnuts during the summer before school started in September, and on weekends, the entire family worked together to get the job done. Dominga González packed a special lunch for all of them and they would make a day of it. Pictured is Esther Pulido, who is married to José Pulido.

From the end of World War I in 1918 to the end of World War II in 1945, Howard Pressey served as manager of Rancho Sespe. Keith Spalding hired him after W.H. Fleet passed away. Several acres of Marsh seedless grapefruit were planted in 1920. They were so successful that several hundred more acres were planted over the next 20 years. The packinghouse in Rancho Sespe processed lemons and grapefruit, whereas the Valencia and navel oranges were shipped to the Mupu Citrus Association in Santa Paula for processing until 1945, when the lemon and grapefruit packinghouse on the ranch was enlarged to handle orange production. Seen here playing in the snow are Gilbert "Mo" Sosa Jr. and Becky Sosa.

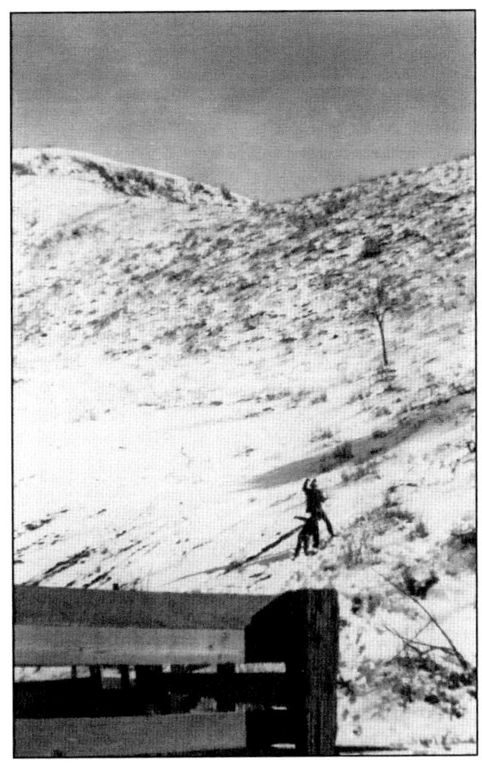

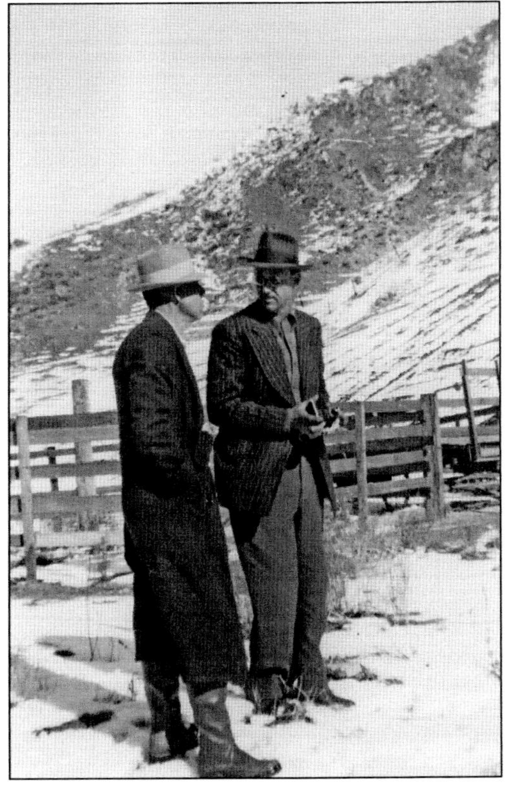

Pictured are Gilbert Lujan (left) and José Pulido. Rancho Sespe was an independent Sunkist packing plant with its own brand of labels. Eudora Spalding chose the labels. A large hotel in Chicago had a rooster as its emblem, and she liked it so much that she had the drawing of a rooster commissioned as part of the Rancho Sespe fruit labels. The top grade of fruit was shipped as Gold Stripe, the second grade of fruit was classified as Silver Stripe, and the lower grade of fruit was Blue Stripe. All the labels had a picture of armor with a white rooster in the middle of it. The diagonal colored stripe across the face of the armor showed the grade of fruit.

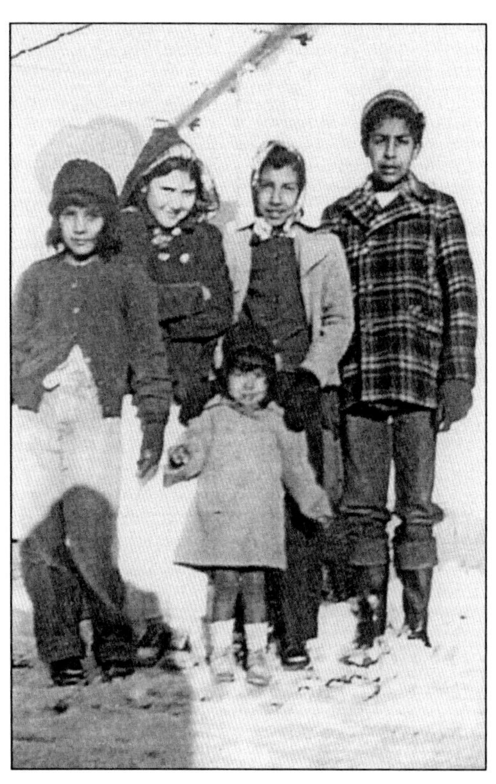

From left to right, Sylvia Castorena, Becky Morales, Laura Castorena, Sally Catorena, and Johnny Castorena enjoy the snow. Laura is pictured below in 1949. She has one brother and three sisters. (Not pictured is sister Carmen Castorena.) The Castorena family lived in house No. 45. She still recalls how much fun they had growing up in Rancho Sespe. It was a small community, and everybody knew everyone else. They were close-knit, and they shared good times and helped one another during difficult times. When a family member passed away, the other Rancho Sespe families took food to the house and attended the funeral. One of the favorite places to go on Halloween was to Mrs. Scott's house, because she would always give the children some great cookies she had baked or huge popcorn balls. Once, they stopped at a newcomer's home and the people had just arrived from Mexico and did not know about Halloween and candy yet. Instead, they gave the children pennies, and the children were very happy about that. (Both, courtesy of Castorena family collection.)

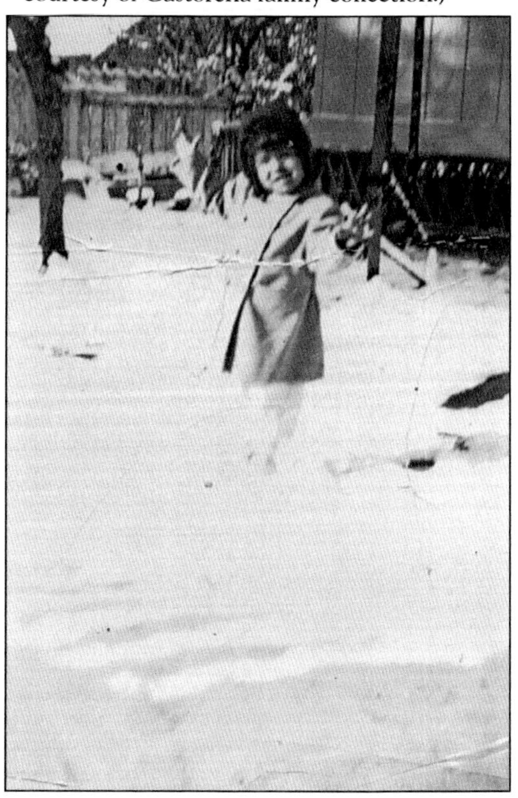

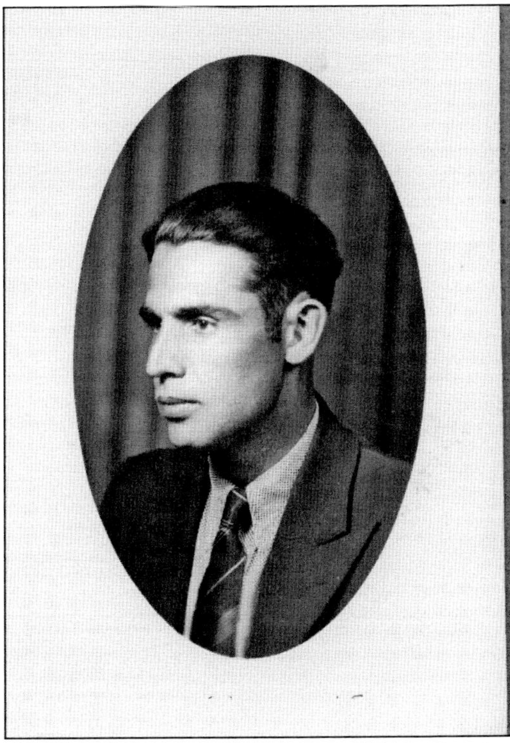

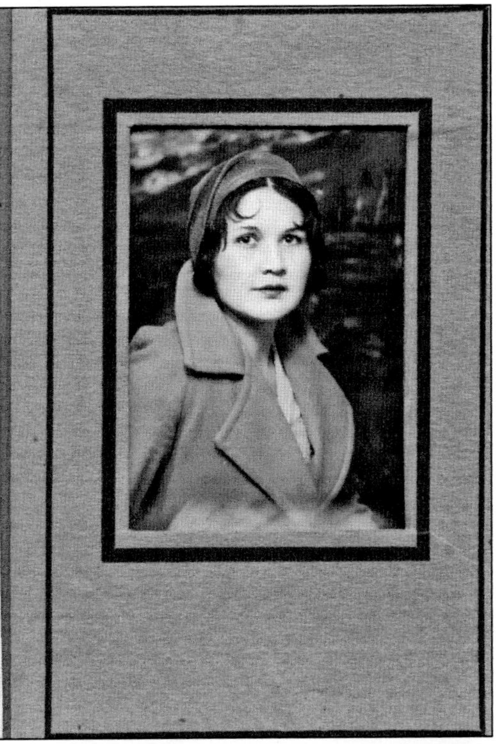

Pictured here are Gilbert Sosa and Alice Sigala. Gilbert met Alice at a dance in Castroville, and they were married on December 5, 1937. This was also Alice's birthday.

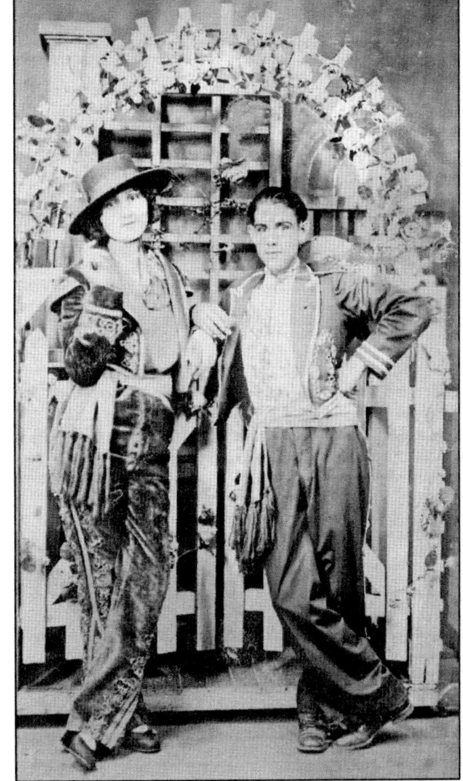

Gilbert Sosa (right) is shown here at age 19. His friend is unidentified. About this time, he was working at Rancho Sespe when the St. Francis Dam gave way in March 1928. He told his family the stories about how he helped others go into the Santa Clara riverbed to recover the dead bodies. During this time, Gilbert traveled back and forth from Rancho Sespe to Salinas to work as he followed the crops and he drove the produce trucks.

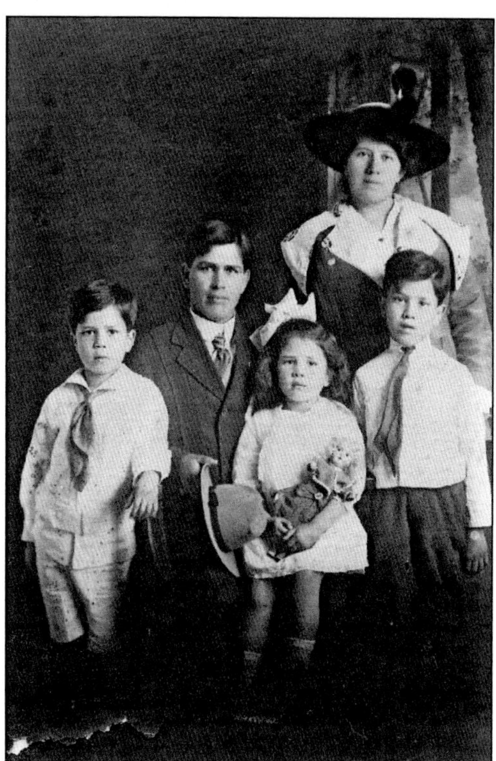

This Sigala family photograph dates to 1918. From left to right are Panfilo Jr., Panfilo Sr., Alice, Minnie, and Tony. Alice was born in Ojai, and she grew up in Ojai, Castroville, and Salinas.

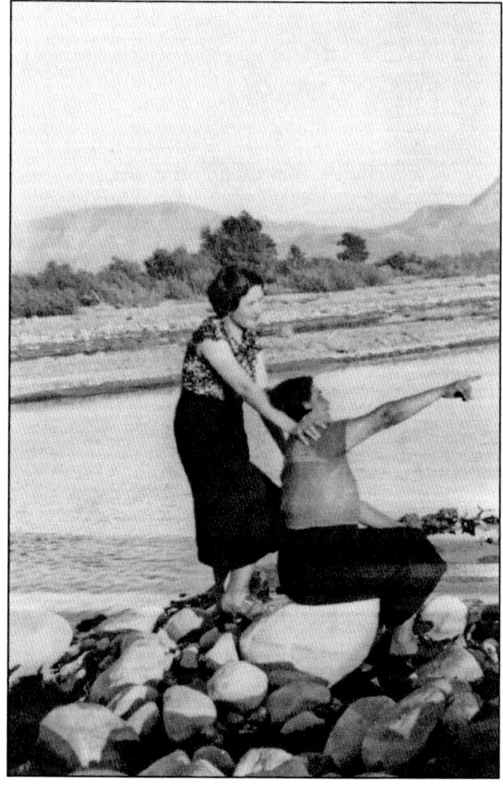

Alice Sigala and Vances Sigala are pictured at the Sespe Creek. The children went swimming in the creek as well as in the Santa Clara River before the California drought days. They would throw rocks and collect some of them, and of course the children collected polliwogs under the watchful eyes of their moms.

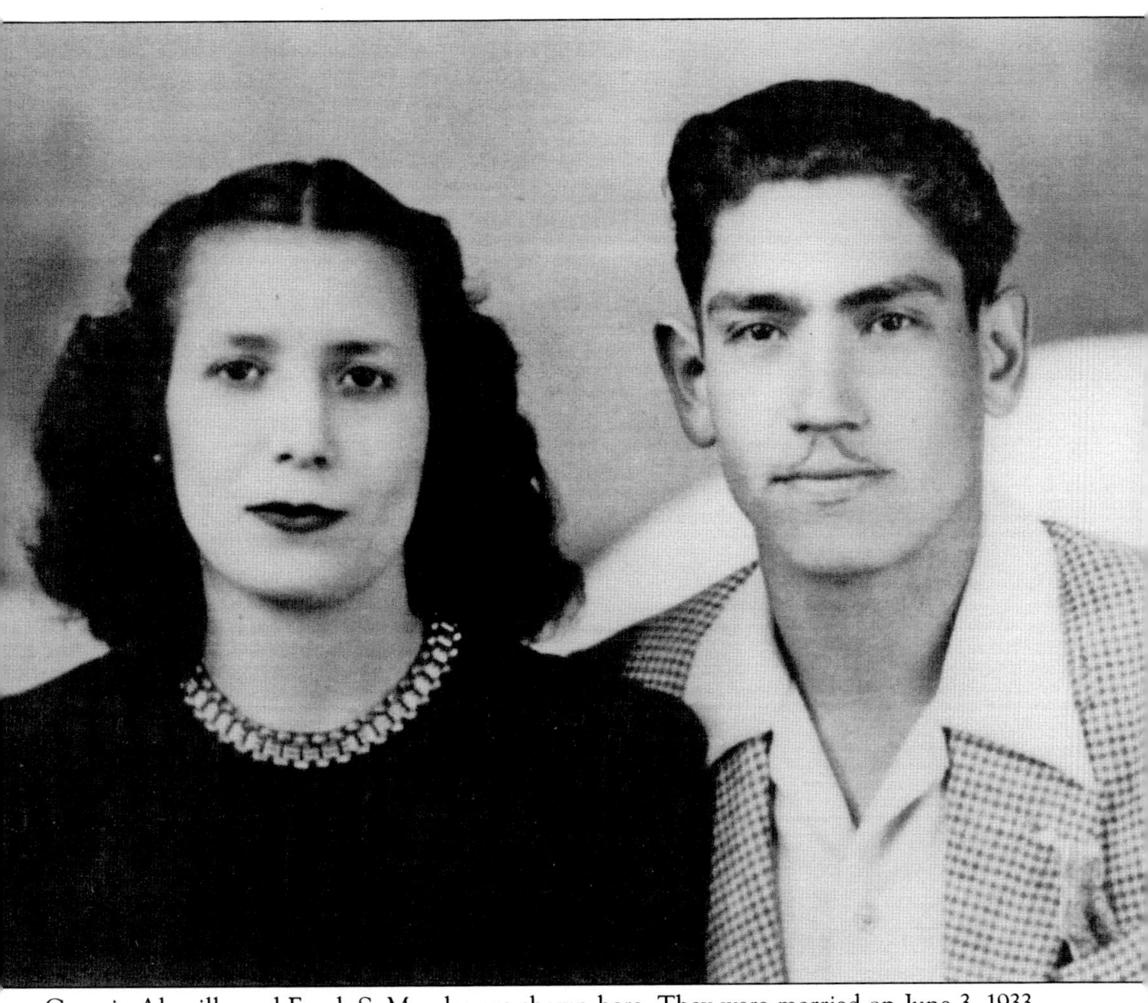

Georgia Alamillo and Frank S. Morales are shown here. They were married on June 3, 1933, in Ventura by the justice of the peace. Frank was 18 and Georgia was 17 at the time. She was originally from Globe, Arizona, and Frank was from Fillmore.

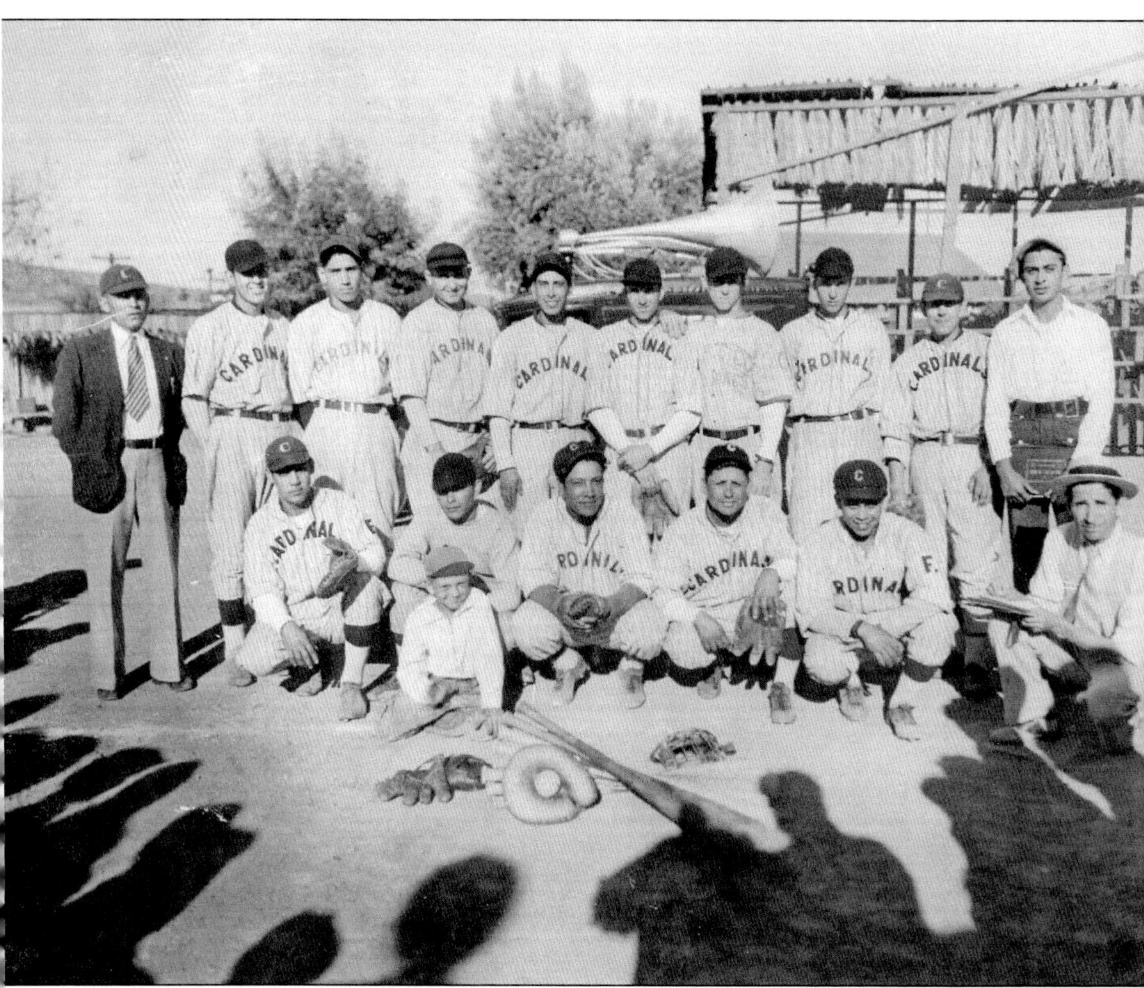

This baseball team photograph shows Frank Morales in the back row, third from the left. Frank played for the Fillmore Merchants as well as other teams in Ventura County, including the Cardinals pictured here. An informal competitive league played from the 1930s through the early 1950s and included teams from Fillmore, Piru, Sespe, Santa Paula, El Rio, Ventura, Oxnard, and Camarillo. The teams played games in Santa Barbara and San Diego. One summer in the early 1940s, the Fillmore Merchants traveled to play in Calexico, Mexico, and the Fillmore team was victorious.

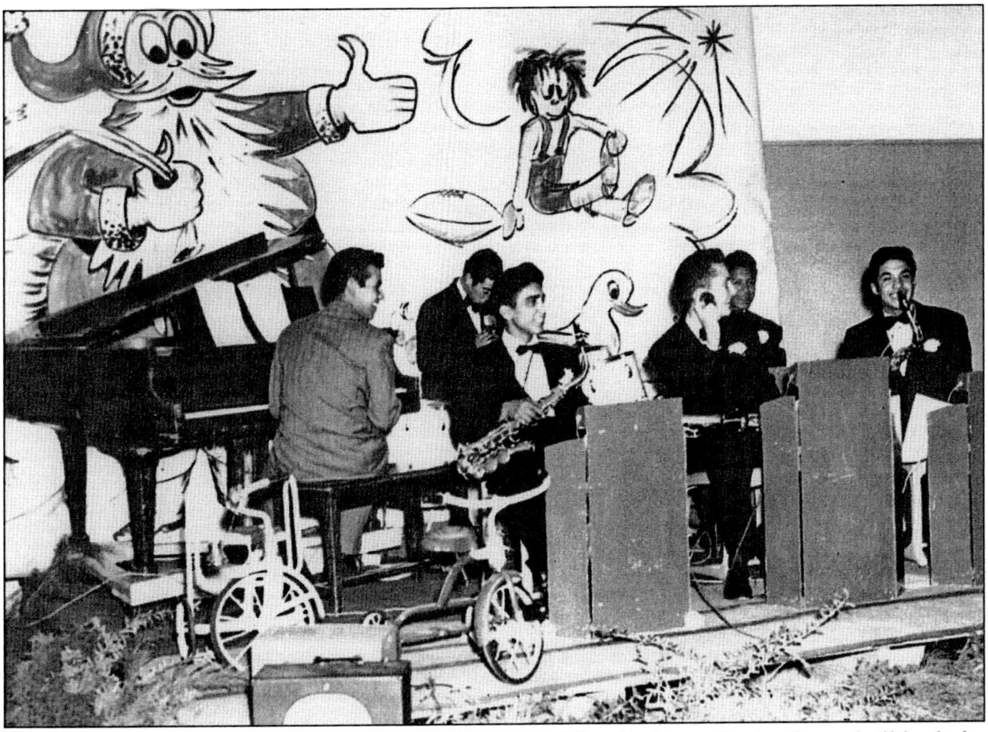

This teenage dance band is playing at Rancho Sespe. The Spaldings had a dance hall built for the people in Sespe Village, and it was here that many festive celebrations were held. Dances with live music were always enjoyed by the residents of Sespe Village. Those identified in this teenage band are Arthur Pulido at the piano, Henry Ramirez (on the left) with his saxophone, Domingo Cardona (at center) at the drums, George Garnica with the saxophone, and Tony Ybarra from Fillmore (in the back row) playing the saxophone.

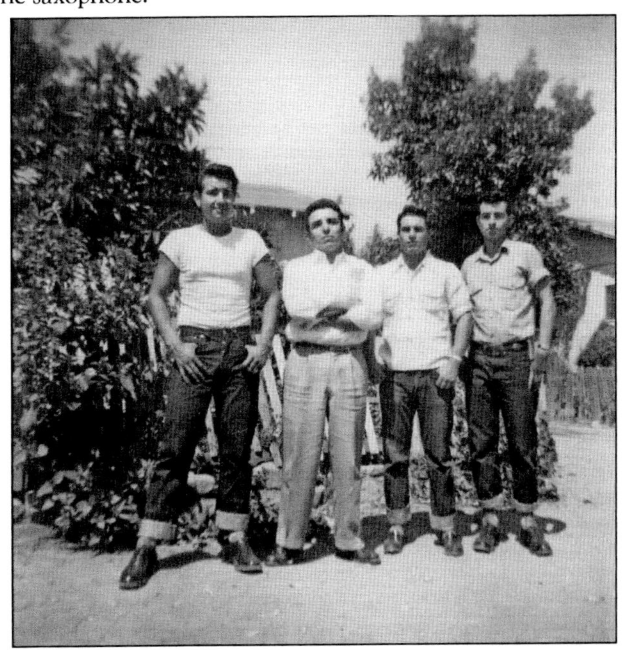

Teen boys would always get together for a good time, though most of the families kept to themselves on the ranch. There was no entertainment to speak of other than playing games with each other in the house or in the yard. Many families had friends in the neighboring village of La Campana and would go there to socialize. They would play guitars and sing, cook, and share feasts and celebrations. The families from La Campana would go to see their friends from Sespe Village the following week, and the friendships were reciprocated.

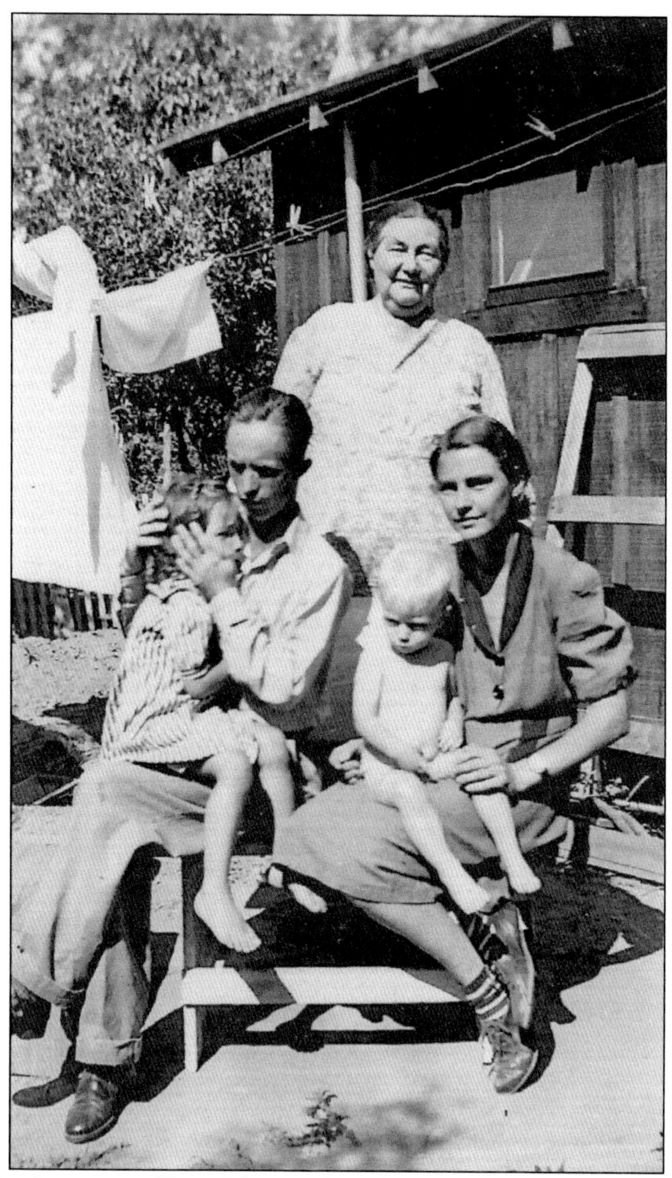

The Crockett family is pictured here. The Crockett children are Oma, Jerry, Joan, Gail, and Dale, and they had a dog, Skippy. They remember as children playing in the stream across the street from their house. There were three acres of bare land where they would also play games. Close to their house were two oak trees they would climb and play Tarzan and Jane. They had placed some old box springs to break their falls, and recall playing there for hours. There was a walnut orchard beyond the ranch houses where they would also play. One afternoon, they discovered a dugout about three feet high, six feet wide, and eight feet long, and they named it "the Fort." They made bows and arrows out of sticks—and they had real arrows too—and they chose teams to "fight" one another. They recall that they had so much fun but did not realize the dangers that lurked in their actions. Once they arrived home, their father asked them what had happened, because they were all covered with marks and scratches. They brought him the bows and arrows, and he broke all of them to pieces and told the children they could not play such a dangerous game.

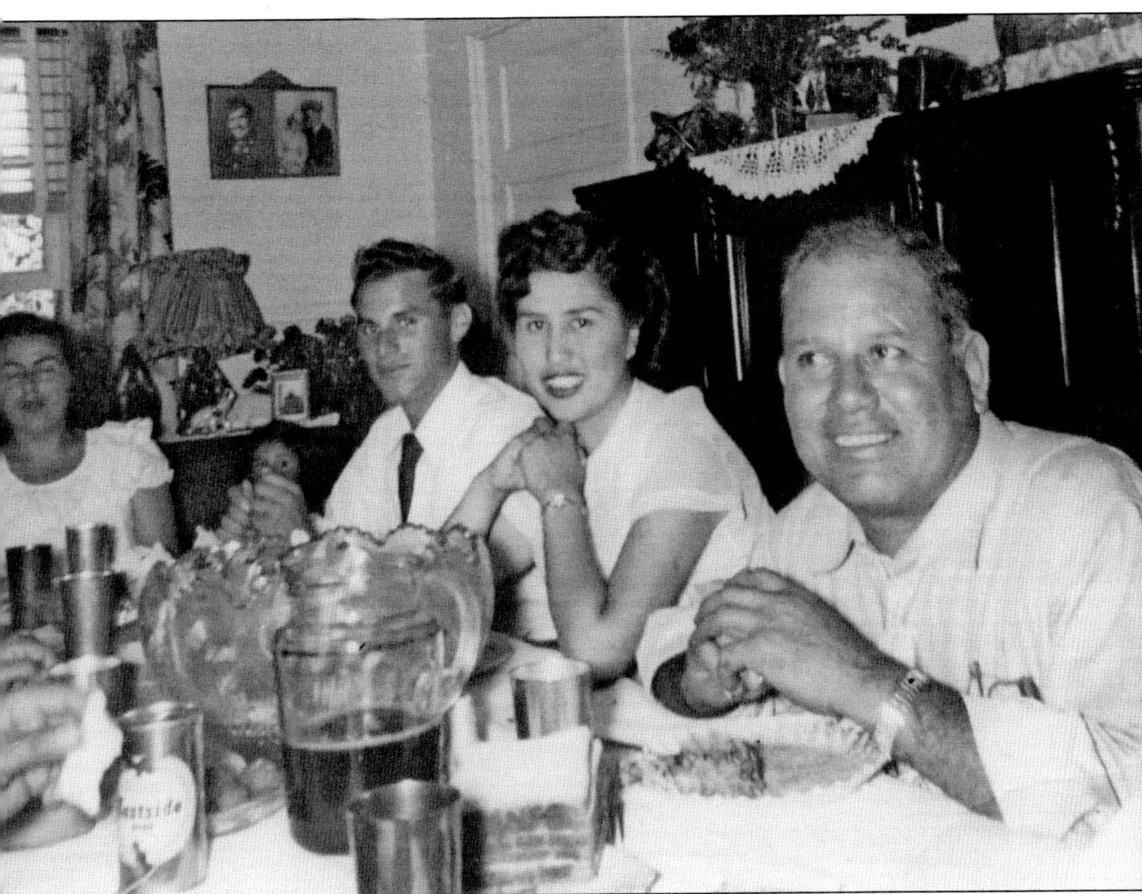

Families would talk around the dinner table. The workers were let out early on Saturdays so that they could go into town and purchase what supplies and food staples they needed. There was no work on Sundays, and all the stores in town were closed except one of the drugstores, which was open for a few hours. Petra González remembers when her father would go out to the highway and catch the Greyhound bus to Santa Paula to go to a market there—El Brillante, owned by Juan Salas. Juan González would normally buy 100-pound sacks of chicken feed, 100-pound sacks of corn for grinding and making tamales, 100-pound sacks of flour for tortillas, pinto beans, and lard. At the end of the day, after Salas closed the store, they loaded up Salas's truck with the purchases. Later, Mr. González acquired his own car and was able to go to the market in Fillmore. Petra González remembers he had a 1929 Buick. He would also drive to Oxnard to purchase lugs of tomatoes for canning. He brought home string beans, spinach, and peaches for jam.

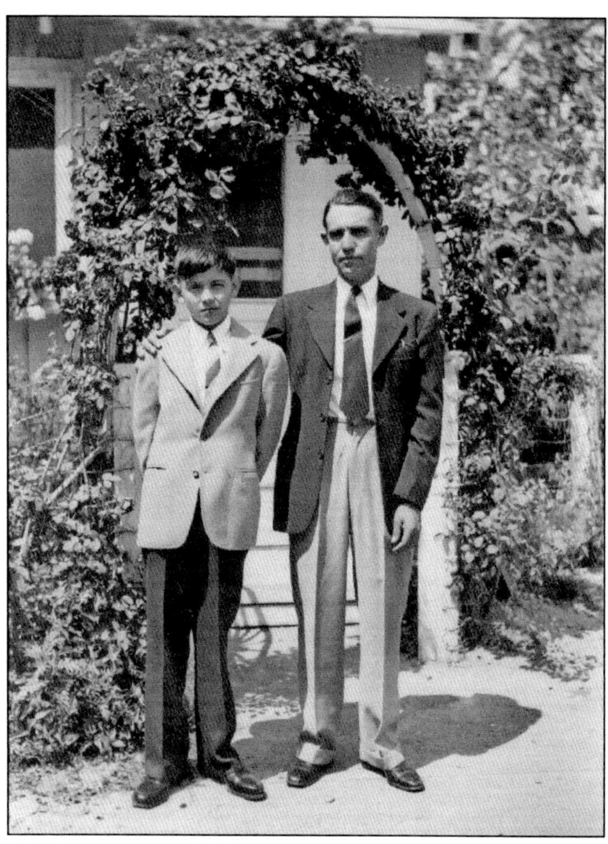

Richard Renteria (left) poses with Gilbert Sosa in front of the Renterias' house on Richard's confirmation day. The families attended mass at the Castorena home when the priest would come to Sespe Village. Later, as more families had their own vehicles, they would drive into Fillmore and attend mass at the St. Francis of Assisi Catholic Church on Central Avenue.

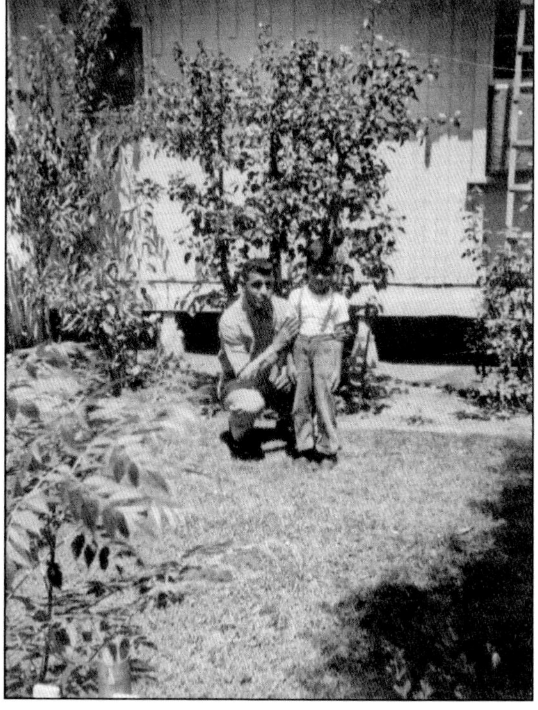

Ernest González is shown with his nephew. There were many employers along the California coast who would hire workers to harvest their fruit but did not provide a place for them to live. Many families who had gone elsewhere to pick fruit lived in tents on the ground with their entire family. Some of these families were too far from the town centers, and their children did not attend school. In the 1930s and 1940s, a group offered union memberships to the workers for better pay and work conditions. Some of the men signed up, and they became ostracized by those who did not join the union. Those who signed up went on strike and picketed to try to prevent the others from reporting to work. However, there were families who lived and worked at Rancho Sespe and did not join the union or go on strike. Instead, they crossed the picket lines, had fruit and other things thrown at them, and were verbally assaulted.

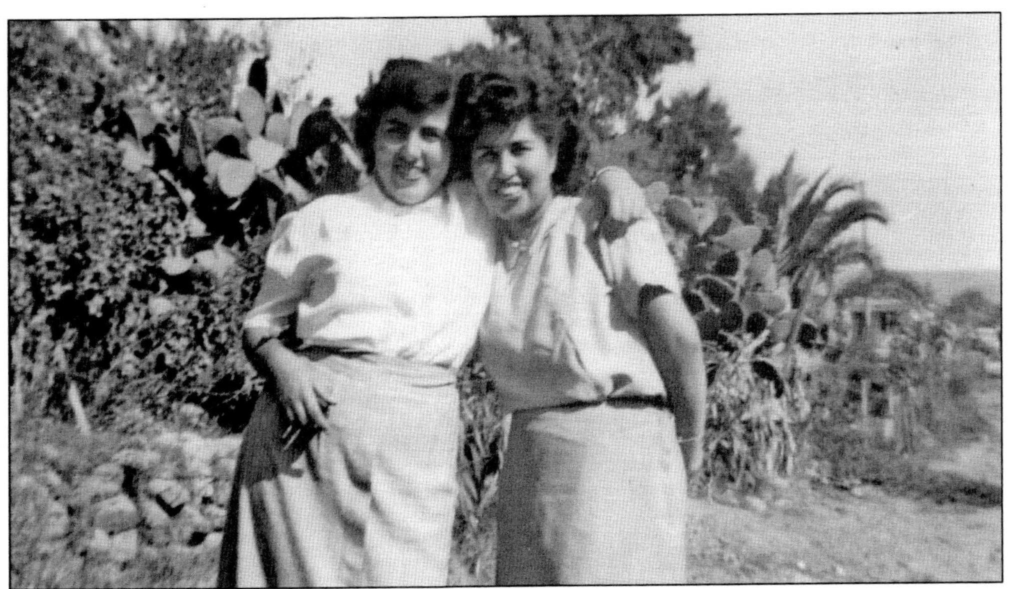

Shown here are Lupe De La Torre (left) and Glafida Sigala. Petra González remembers how the merchant Mike (Mik-ay) would come to the ranch every month. He had men's work pants, T-shirts, material, thread, trim for clothing and sewing, dry goods, utensils, and various household items. He would extend credit whenever the women needed something and could not pay him until the following month. There was a market owner in Fillmore who would bring fresh beef, vegetables, and other things not easily available at the ranch. A milkman delivered milk, butter, and eggs. Later, the Sanitary Dairy made deliveries to Sespe Village. All of this was during the Depression, and wages for the men were as little as $50–$75 per month. The families did not have to pay rent, but $4 was taken out of their paychecks for utilities. Many families learned to cut their sons' hair, the moms sewed their children's clothes for school, and others knew how to put half soles on shoes.

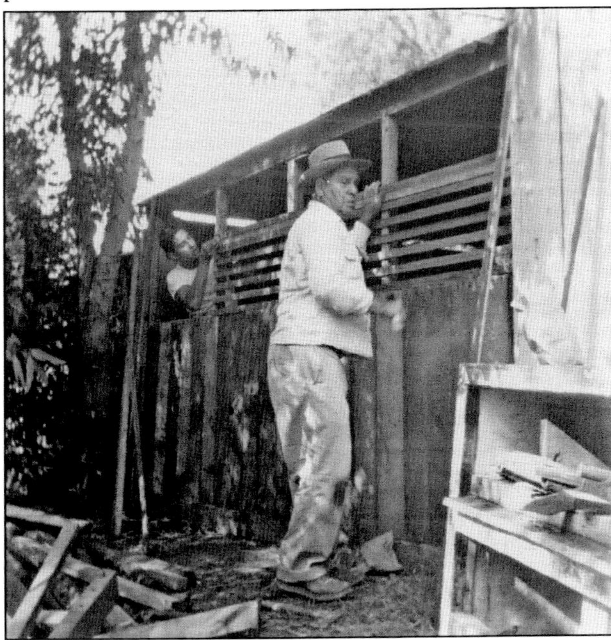

"Grandpa" Panfilo Sigala was a carpenter at Rancho Sespe. Panfilo Sigala was Alice Sigala's father and Becky and Gilbert Jr.'s grandfather. He enjoyed fixing things, and those who lived in Sespe Village would call upon him when they needed help with a project.

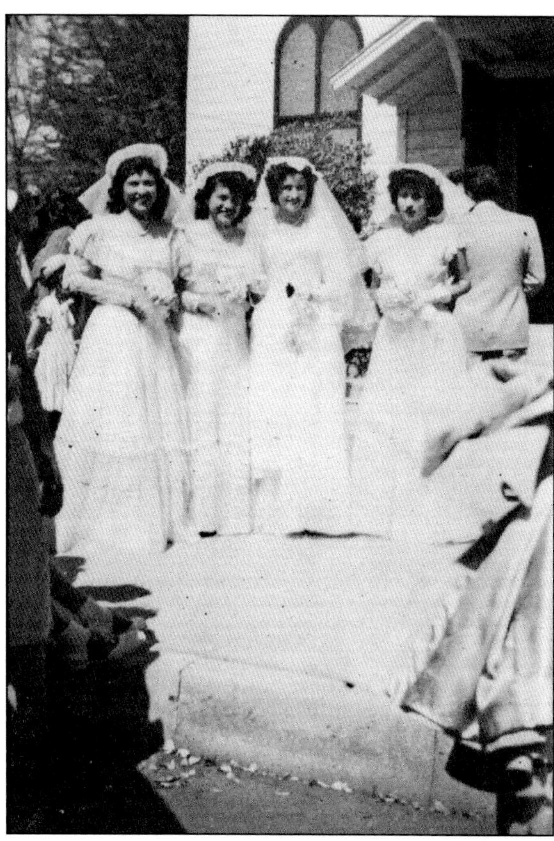

Rosalinda Renteria is pictured on her wedding day. She was married at St. Francis of Assisi Catholic Church in Fillmore.

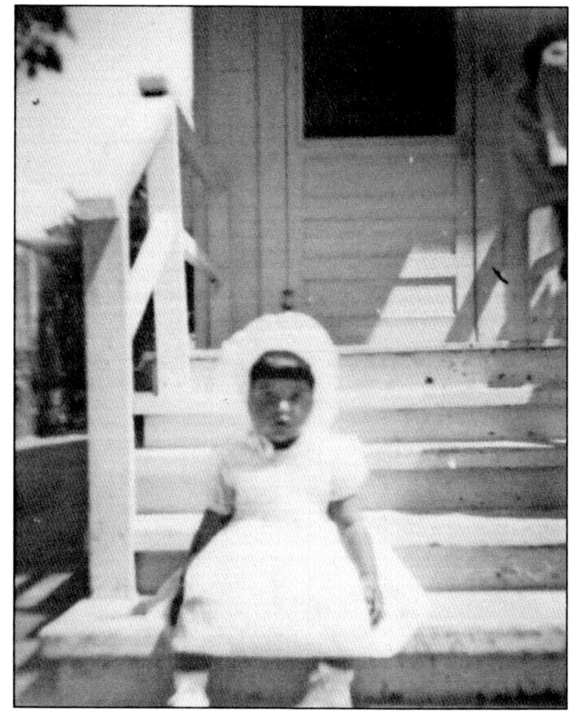

Alex and Carmen Davis's daughter Sandra Davis is sitting on the front steps at Becky Sosa's house in Rancho Sespe. The houses were neat and clean, and everyone in the family had chores. All the houses had flowers growing in the front yards. The homes were all painted white in the 1930s, and plumbing with drainage was installed and screened porches added for the new water heaters and laundry tubs. The women no longer had to wash clothes outside in the yard. Bedrooms were added to the houses as well. In the beginning, the houses had one bedroom, a long living room that was also used as sleeping quarters, and a kitchen. The expansions were a welcome addition for comfort.

Federico González is shown at Ventura Beach. He and his wife worked for the Spaldings. Federico and his wife had five sons and four daughters. All five sons served in the military during World War II and the Korean War.

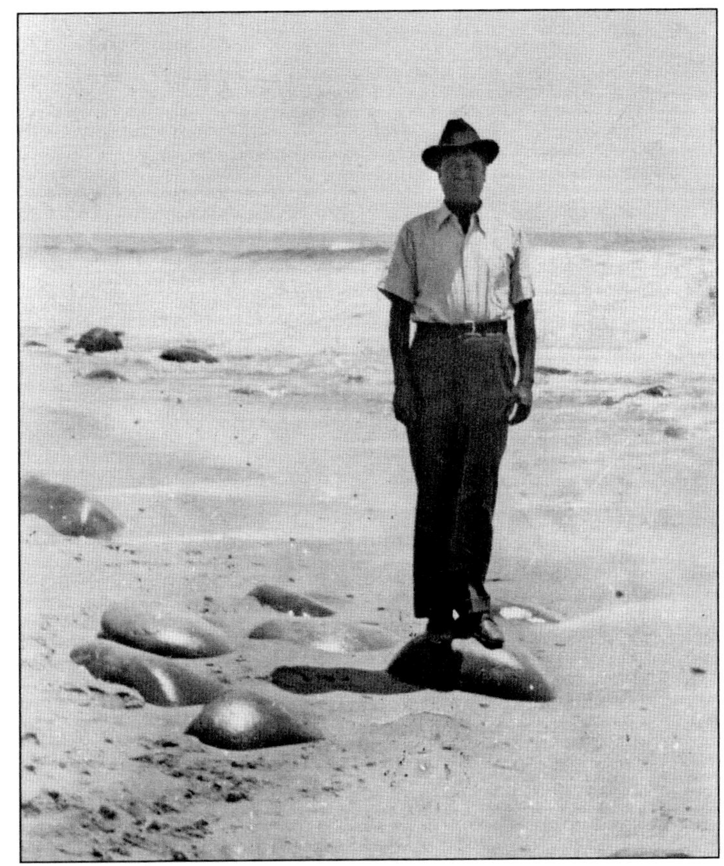

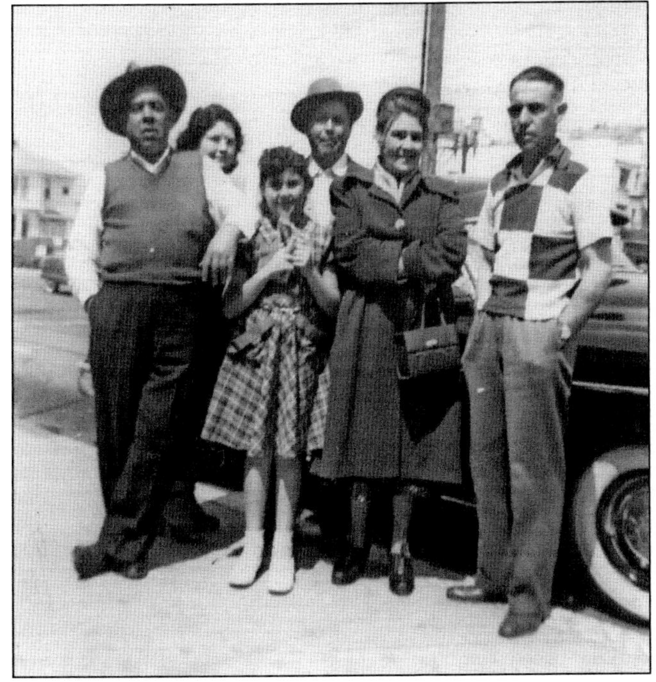

The Sosa and Sigala families enjoyed their Sunday drives after church. Here, from left to right, Tom Valenzuela, Alica Sosa, Becky Sosa, Panfilo Sigala, Mary Valenzuela, and Gilbert Lujan Sosa are preparing to take out the new 1949 Chevrolet.

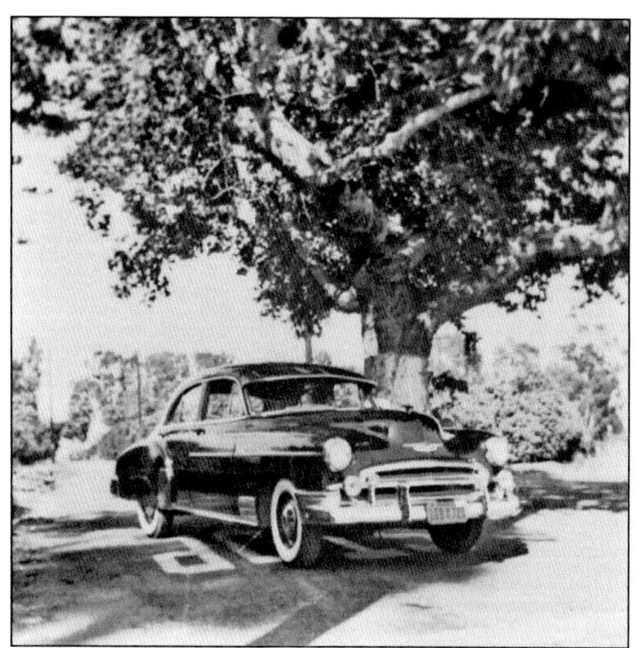

On Sundays, Perry Crockett took his girls out driving in his own car. This was before driver's education classes were offered. Oma and her friends took turns at the wheel. Perry was very calm and at ease as a driving instructor, and they never met a highway patrolman on the road. When Jerry Crockett was 13, he decided he wanted to learn to drive too, and Perry, his father, took him out one afternoon. As Jerry was driving out of the ranch, a highway patrolman pulled up beside them. Perry told the patrolman he could not believe he did not have the right to teach his son how to drive! The patrolman let him go and did not ticket him.

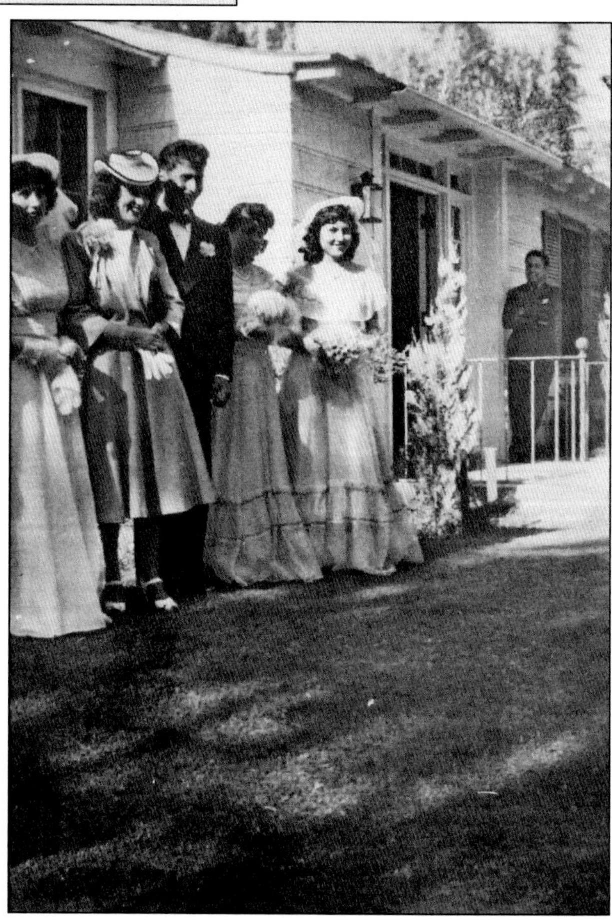

Rosalinda and Ruben Preciado have their wedding reception at the Ebell Clubhouse in Fillmore. The Ebell Clubhouse is designated as a Ventura County Historical Landmark.

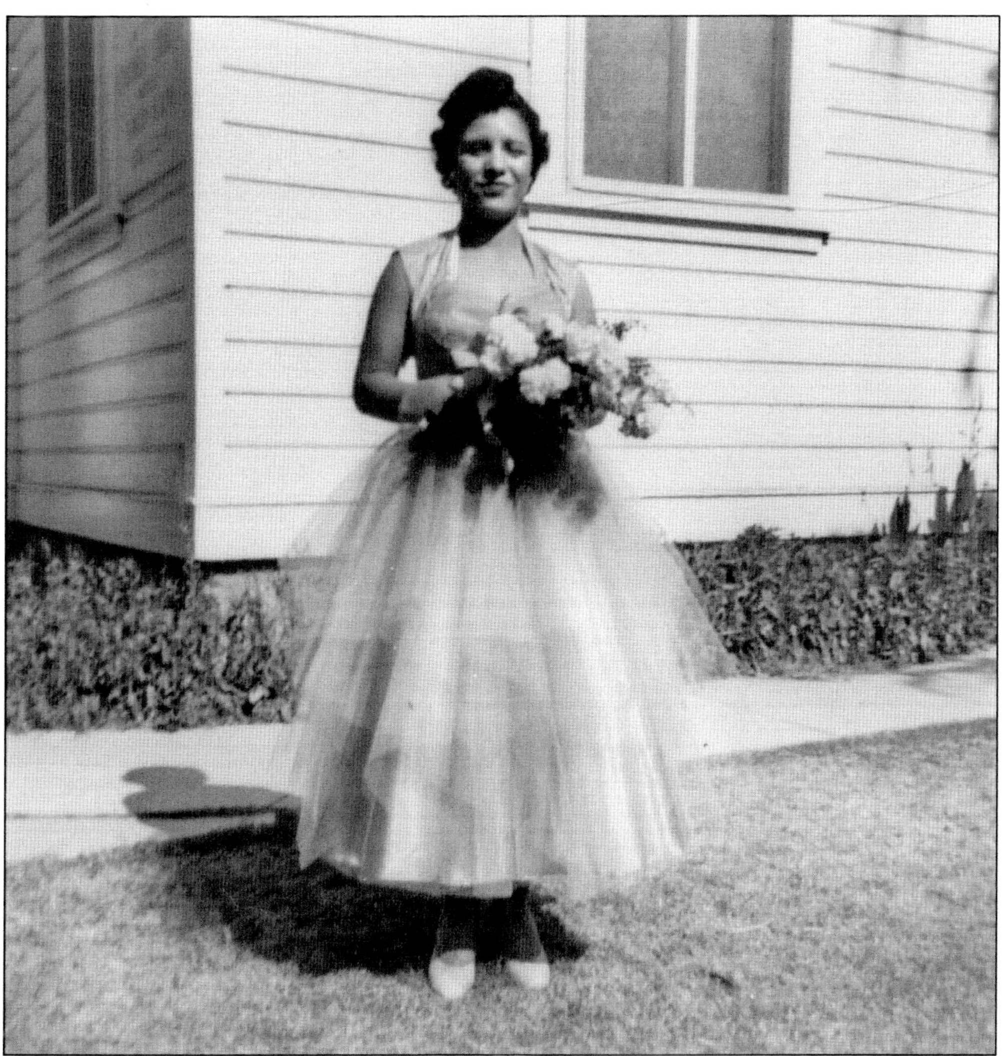

Carmen Castorena is shown here as a bridesmaid for Rachel Ramirez. Many workers starting families at Rancho Sespe knew that they could not do any better anywhere else. The ranch had one owner, the Spalding family, and it was not an association; they had housing provided for them, and each home had its own garden. Petra's father, Juan González (no relation to Ernest González), shared with his family how he had overheard Eudora Spalding talking to a salesman who wanted to sell her some automated equipment that would do jobs around the ranch property. This equipment could pick up rocks mechanically to be used for building rock walls around the orchards. She said no, because her workers would not be able to mark time in order to make a living if machines did all the work for them. The Spaldings did not allow the union to come in and organize the workers. If a laborer signed up to join the union, he could not report to work, and he and his family had to leave their home. Many families were displaced by this event. Driving into Santa Paula on the outskirts of town, one would see fields covered with tents. It was a terrible hardship for those families to survive.

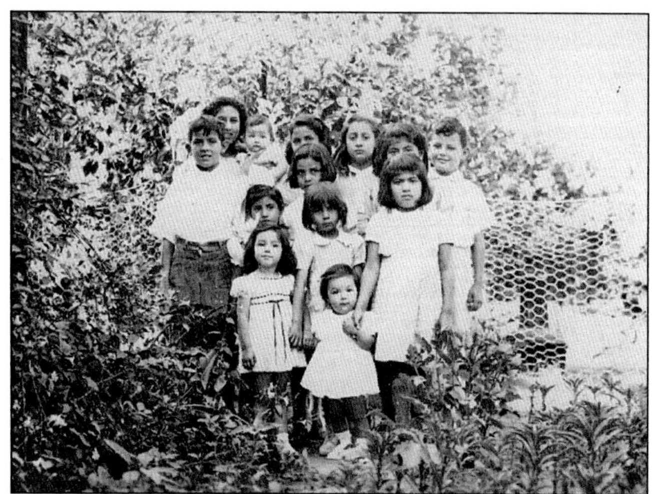

Some children gather for playtime. The children played kick-the-can, and they would swing on the rope tied to a tall tree in Sespe Village. One of their favorite things to do was to run down to the riverbed and play along the edge and throw stones or gather polliwogs. Identified here are Richard Renteria (far left, second from top) and Irene Villanueva (far right, second from bottom. (Courtesy of Robert Villanueva family collection.)

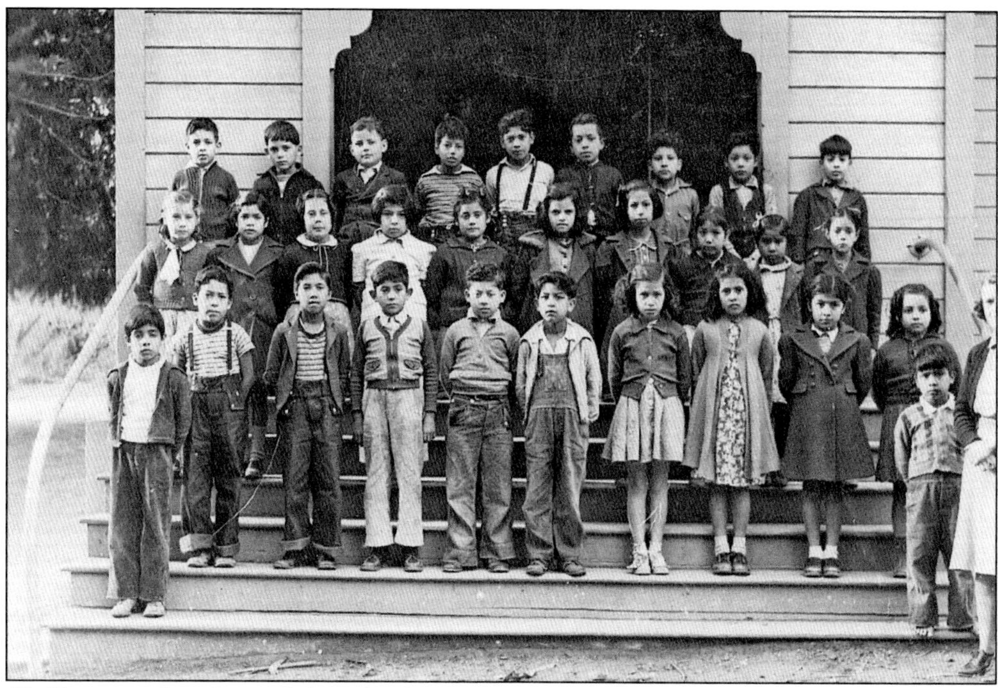

Children stand in front of the school on Seventh Street in Rancho Sespe. Henry Ramirez remembers growing up in Rancho Sespe from 1930 to 1960. It was a good experience for him. At age six, he attended kindergarten at the small schoolhouse in Sespe Village, and Martha Stoll was the teacher. The school was a small house behind the dance hall. Esther, Henry's older sister, would walk him to school, rain or shine. Henry started first grade at the school on Seventh Street, close to Kenney Grove Park. In the days before busing, the children would walk two and a half miles every day, except on rainy days. He recalls the teachers there were a Mrs. Hiller for first grade, Dorothy Rykeman for second grade, and Gertrude Campbell for third grade; she was also the principal. Those identified in this photograph are Pauline Casas (second row, third from left), Robert González (third row, fifth from left), John González and Daniel Perez (first row, third and fourth from left), and Henry Ramirez (first row, far right).

Four

Childhood Memories

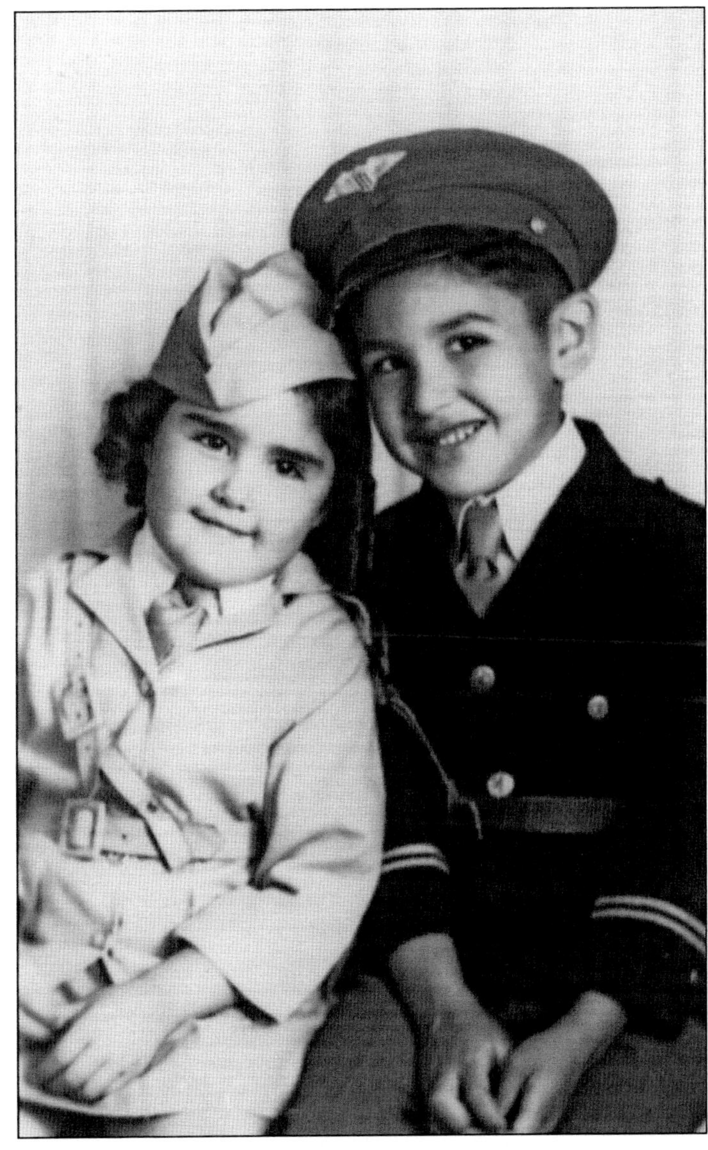

Becky Sosa poses with her brother, Gilbert "Mo" Sosa Jr. During the summer, the children played games such as marbles and bottle tops and jumped rope, and boys rode homemade go-karts. The girls also played hopscotch and jacks. Traditional play such as playing with dolls, playing house, singing, and backyard sleepovers were also common. The children loved running free and going down to Sespe Creek and the Santa Clara River to wade and swim on a hot summer day. They collected polliwogs as well as frogs, pretty rocks, and wildflowers.

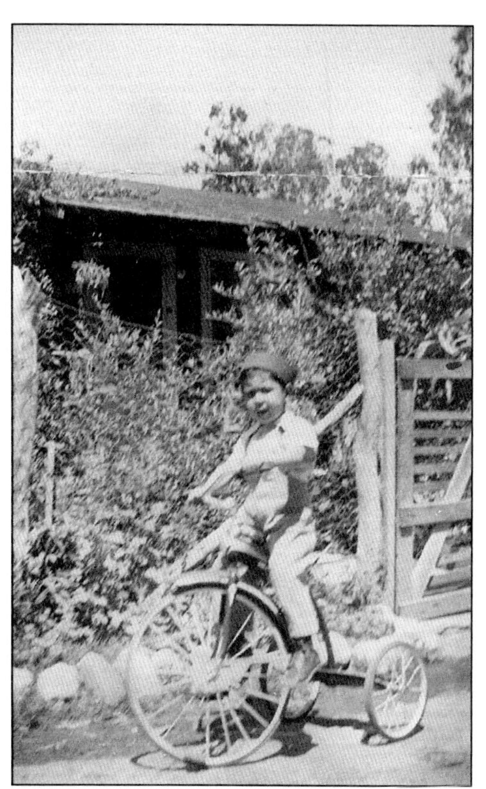

Henry Ramirez recalls that one Halloween the boys got together and collected some large trash cans, and they waited for cars to come up the hill below house No. 129. When they saw headlights rolling towards them, they pushed the trash cans down the hill as the cars were coming up. Gravity took care of the rest. Of course, the boys ran and hid. That was the worst thing they did. All the children knew that the moms were in agreement as far as making the children behave; the mothers had the right to spank if necessary, and the children knew it. Bobby Villanueva (shown here) and "Shorty" Williams went to ride their bicycles on the Spalding's golf course. They started doing wheelies on the greens and were caught the second time they went out there. The boys were scolded and never again did they ride their bikes on the golf course. (Courtesy of Robert Villanueva family collection.)

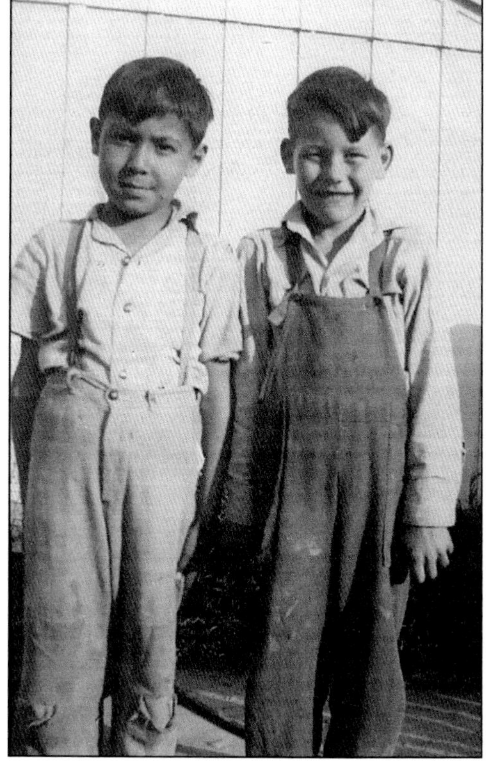

Bobby Villanueva (left) is seen here with his friend Marvin Kruse. The boys grew up in Rancho Sespe, and they have many memories of how much fun they had playing together and with all the other children. They made go-karts that would go fast, although they did not have any brakes, and they had fun rolling down the streets in Sespe Village. (Courtesy of Robert Villanueva family collection.)

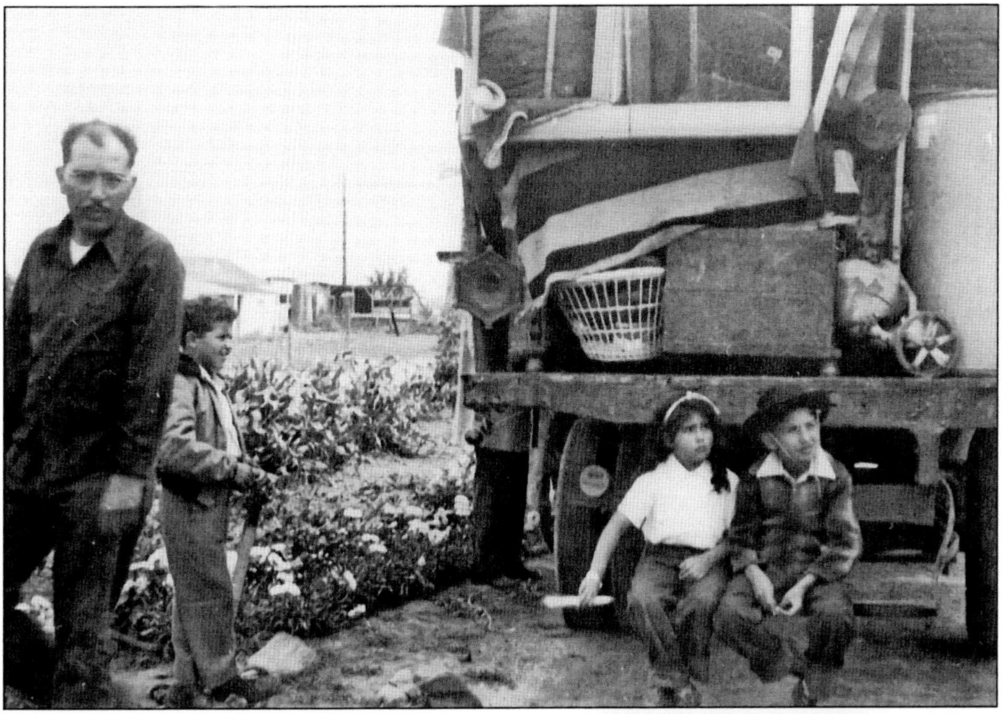

The Sosa family returns to Rancho Sespe from Salinas. Henry Ramirez remembers when the Sosa family first moved away. He and others carried chairs and other small things to the Sosas' truck. Everyone in Sespe Village missed them. The children missed Becky and Gilbert Jr., and the moms missed Alice Sosa. Henry's mom, Guadalupe Ramirez, and Alice were good friends. Henry remembers his mom crying after they left. Everyone was very happy when the Sosa family later returned to Rancho Sespe. They moved into house No. 129. Between two barrancas, this was known as the house with the dip or the "Dip House." "Mo" Sosa was a little older by then, and all the young boys were happy to have him back.

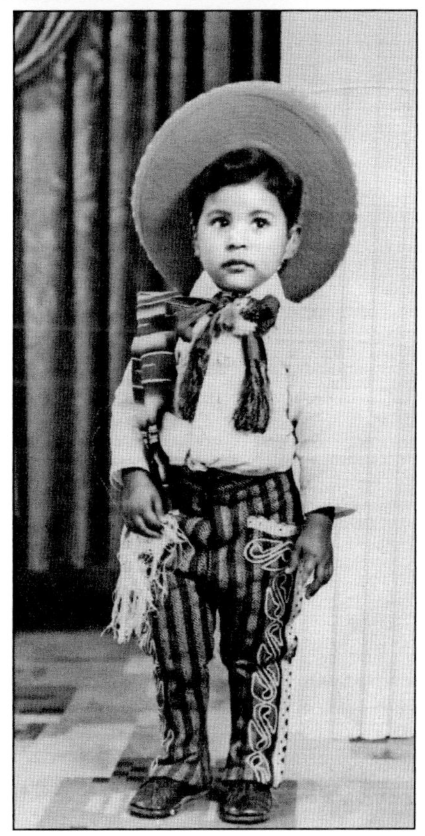

Tomás Torres won first place at the Fillmore May Festival in 1940. Many adults who lived in Rancho Sespe in their youth share their fond memories. Gloria Fernandez, daughter of Tony and Heleen Fernandez, explains that many close friendships were formed. The children were bused to Mountain View School beginning in first grade in later times. They walked to Highway 126 to wait for the school bus. One day, Gloria's father, Tony, had a surprise for her when she arrived home from school. He had adopted a dog for her from the animal shelter. Gloria named him King, and he used to walk the children to the bus stop and greet them when they arrived after school every day.

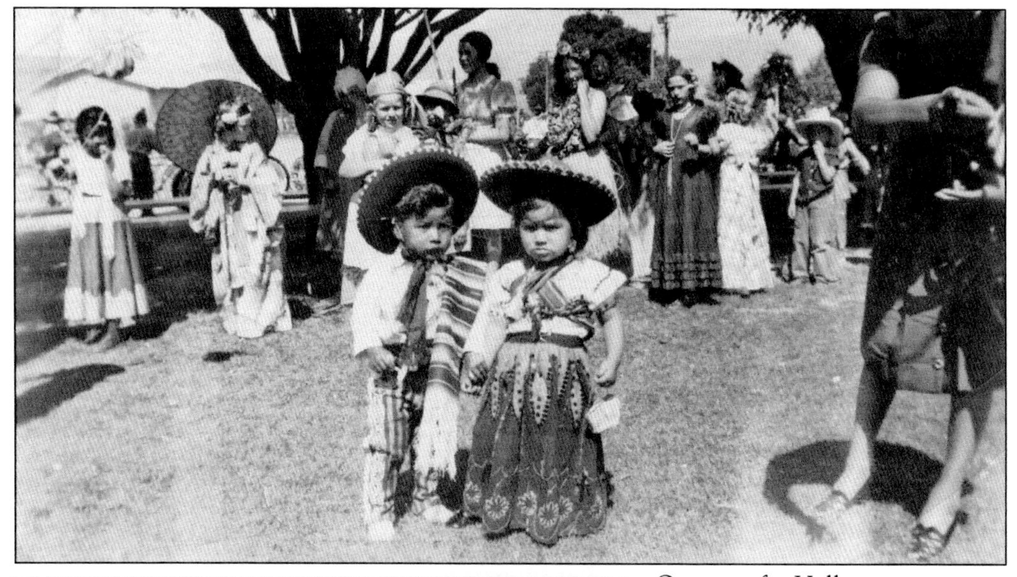

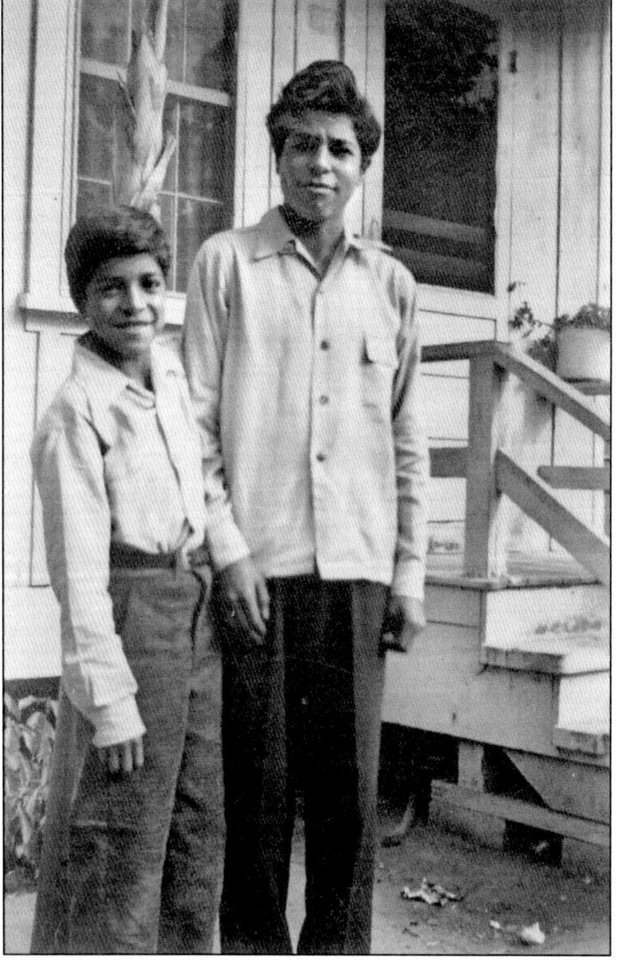

One year for Halloween, some could not afford candy, so the families met and planned a surprise so that not even one child would be left out. The parents of one family made popcorn balls, and they were said to be the best. The time arrived when the men rose to light the smudge pots, and the wives kept the coffee going for the men. The men smudged most of the night. The gentleman who sold vegetables was a very nice fellow; sometimes, he would not charge for them. Many agree that they were all an amazing community, and everyone helped everyone else.

John "Shorty" González (left) and George "Rudy" González are two of the five sons of Federico and Felicitas González.

Becky Sosa is shown at age 12 in front of her house, No. 129. Becky's best friend was Mary Ann Renteria, and the girls would play with their dolls by pushing them in their doll carriages, or they would play *comadres* and pretend to go shopping.

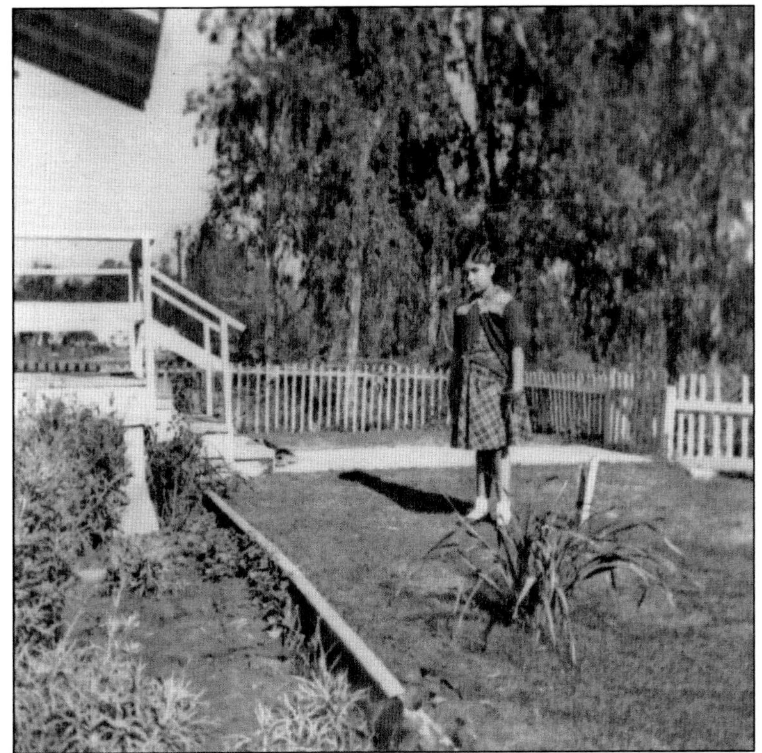

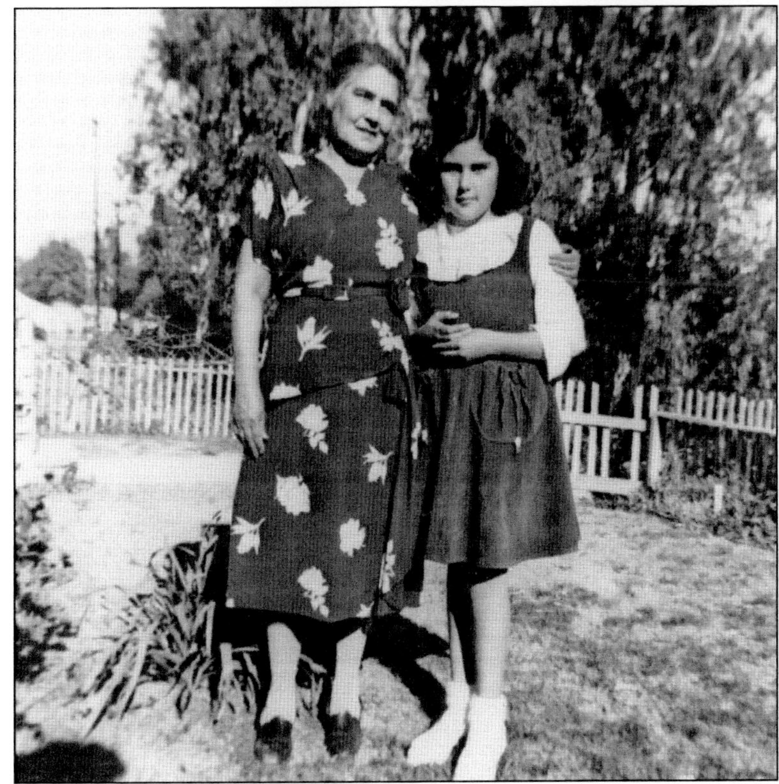

Maria Valenzuela (left) is pictured here with Becky Sosa. Becky is wearing her favorite red corduroy jumper that her mom made for her.

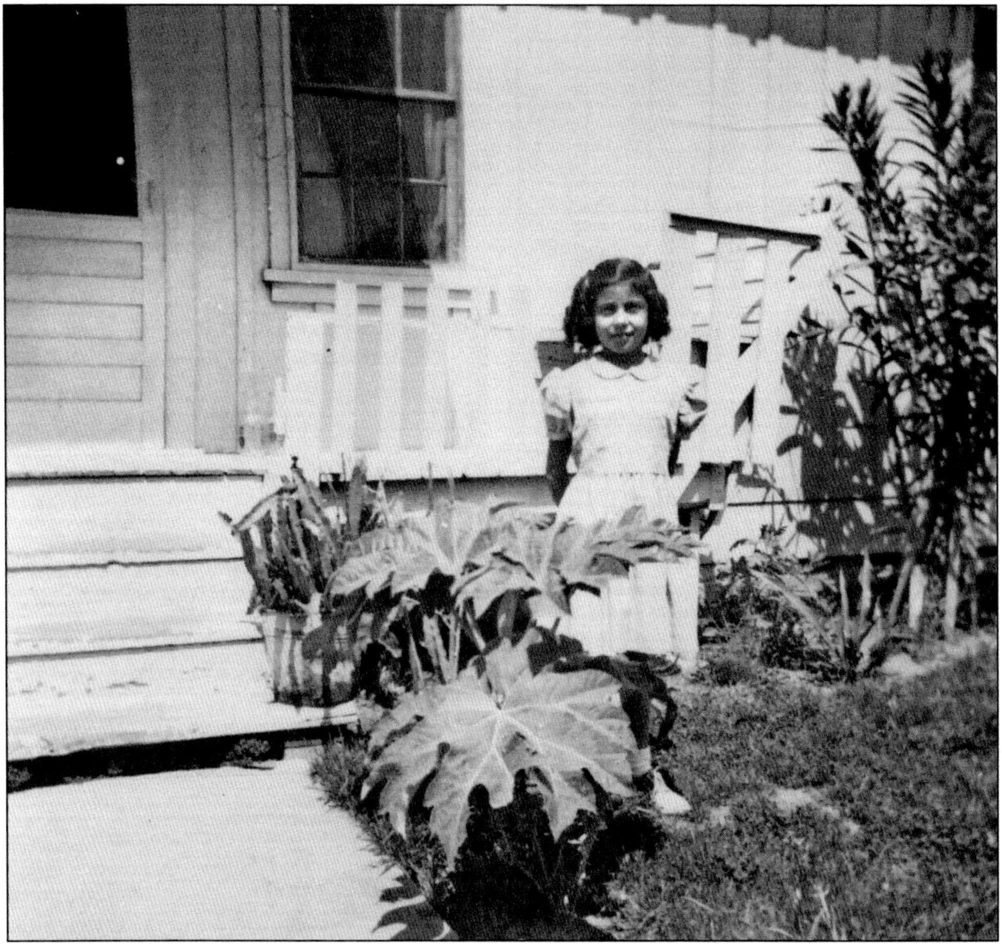

Mary Ann Renteria (pictured) and Becky Sosa would climb in the family car and pretend to drive to the grocery store or into town to go shopping for their dolls. They sat in the front seat of the car and talked as they were "driving into town"—they played this for hours at a time.

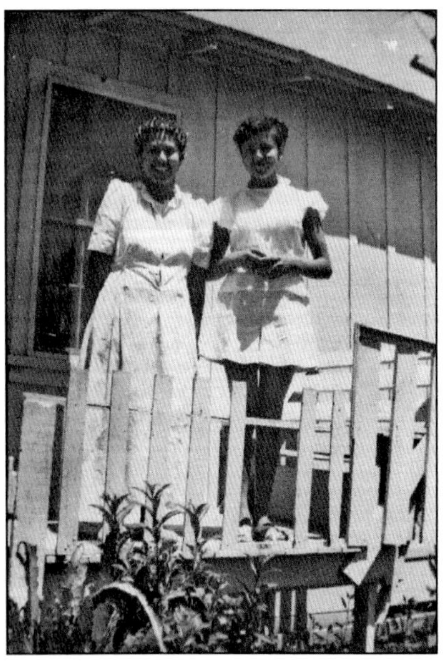

Glafida Sigala (left) and Amparo "Amy" Renteria pose in front of Amy's house, No. 124. Glafida is Becky Sosa's aunt, and she lived with the family for a while. Amy is Mary Ann Renteria's sister.

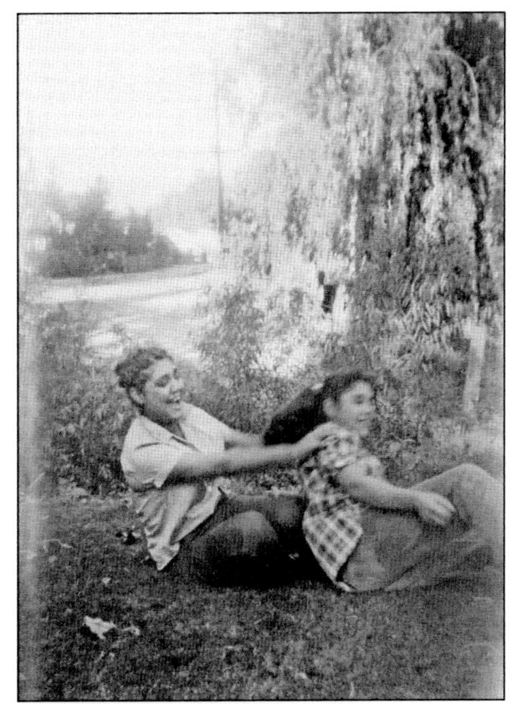

Angie Rodriguez (left) and Becky Sosa are shown here. Kris Jaynes Hyatt recalls when her father, Ted Jaynes, used to deliver Prosser's bread to Rancho Sespe. Kris and her sister Kathleen enjoyed riding in the truck when he would go on his bread route. Linda Starnes Faris recalls how she and her sister, Kathy, enjoyed riding in the Sanitary Dairy Truck once in a while. They would accompany their father, Jack Starnes, on his delivery route to Rancho Sespe.

Becky's dog was named Tootsie. The family enjoyed Tootsie, and Becky and Gilbert Jr. always played with her. She quickly became part of the family.

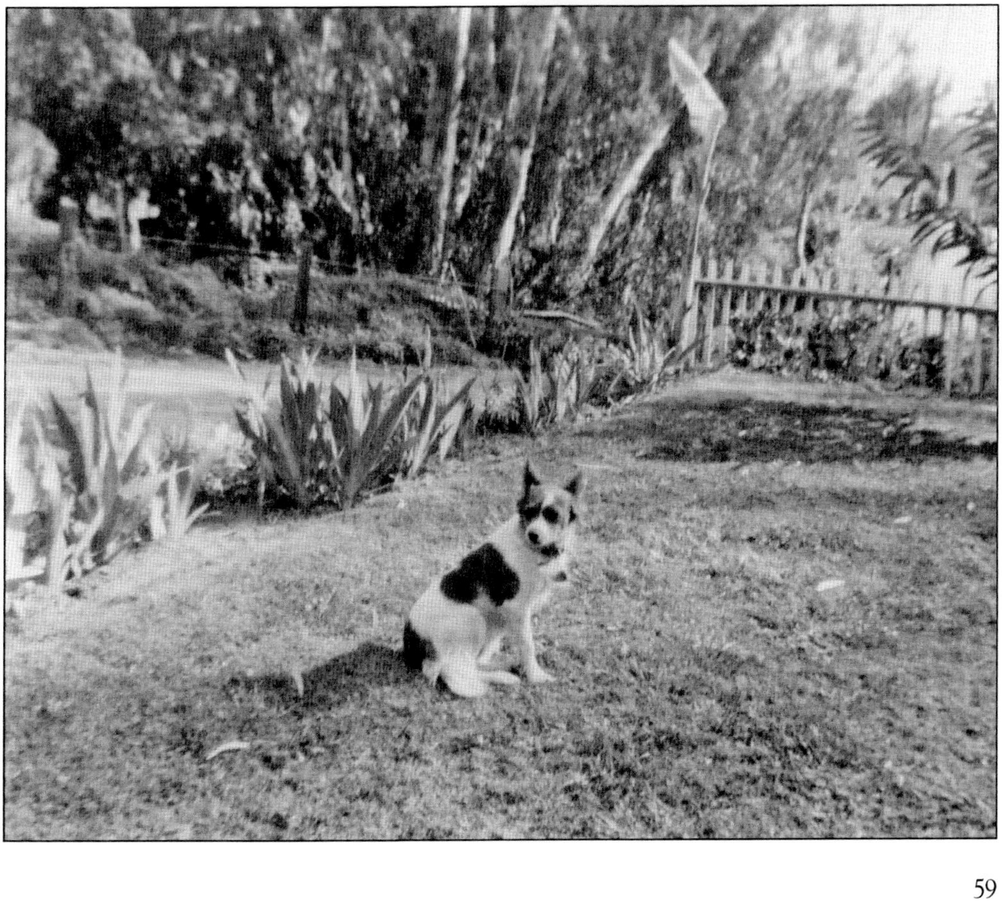

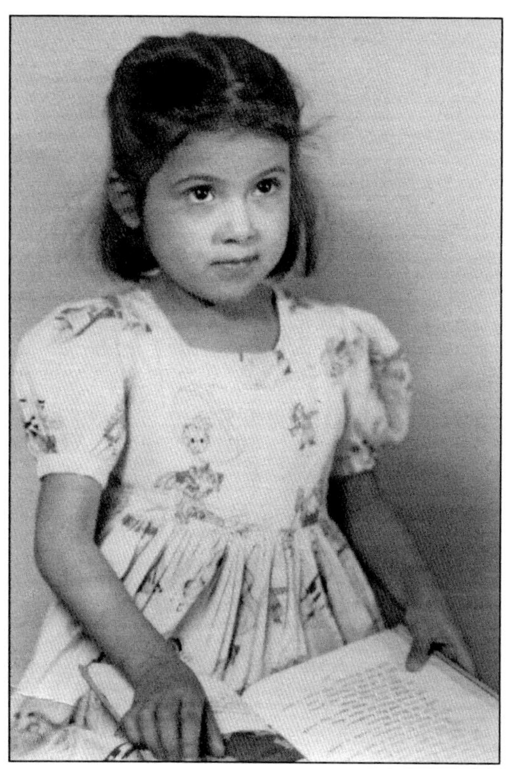

Mary Ann Renteria is pictured when she was four years old. She enjoyed playing with her dolls at a young age. Mary Ann and Becky Sosa were best friends.

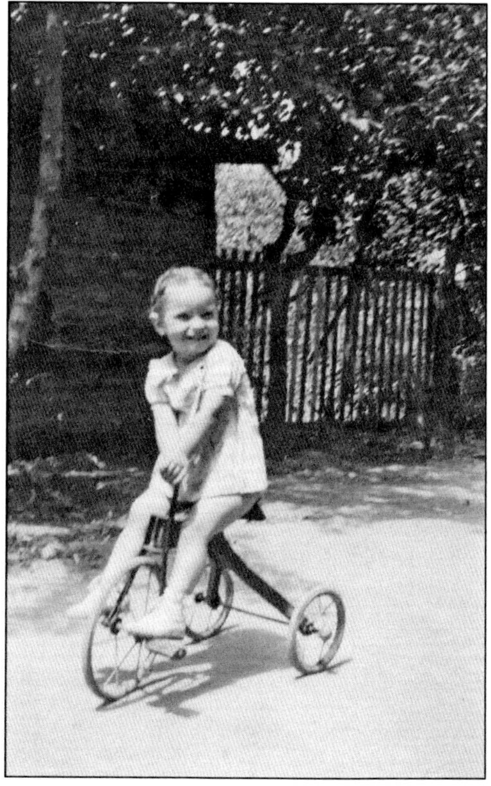

Oma Crockett is shown at age two. The Crocketts lived in house No. 84. The Bragg family lived next door to the Crocketts, and one night the Braggs' house caught fire. They quickly escaped, but in all their panic, they forgot the baby, Sonny Bragg. He was about two years old at the time. Opal Crockett, Oma's mother, realized the baby was missing, so she ran into the burning house and grabbed Sonny out of his crib, saving his life. Opal became the heroine of the ranch. (Courtesy of Gail Crockett family collection.)

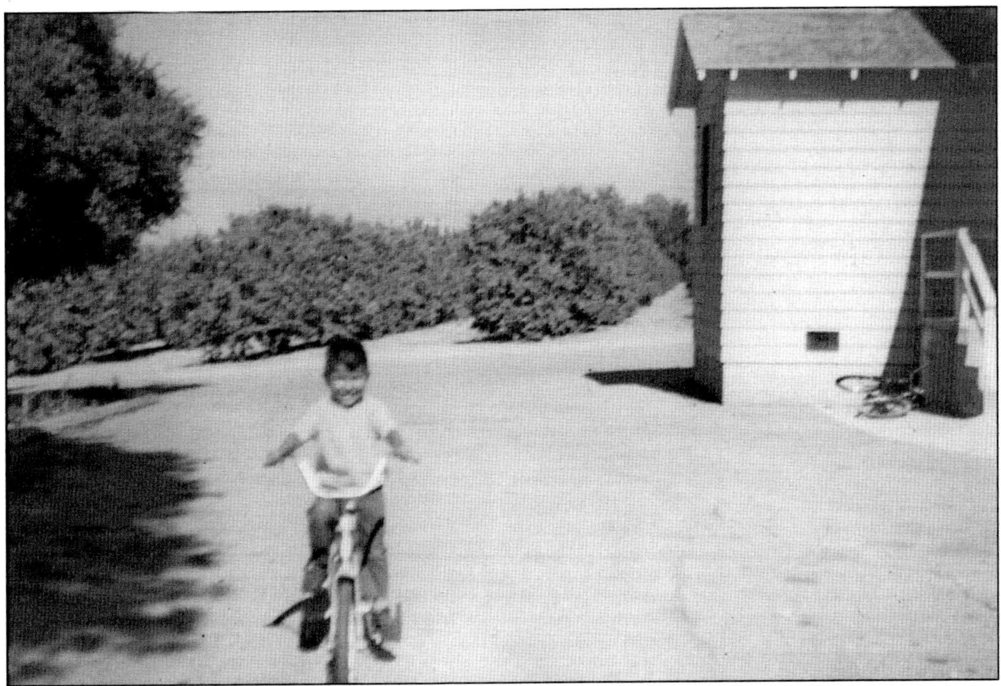
Tomás Torres Jr. rides his bicycle on the Spalding property. He lived with his family in the Spalding guesthouse.

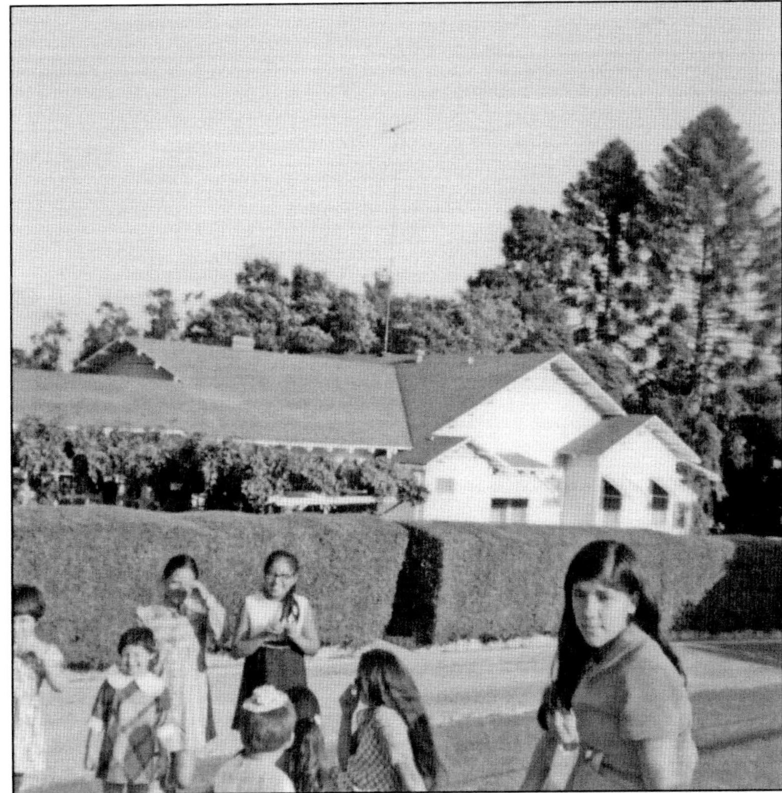
These children have stopped playing to pose for a picture. They enjoyed playing all around the Spalding property. They rode their bicycles, threw balls, and played hide-and-seek.

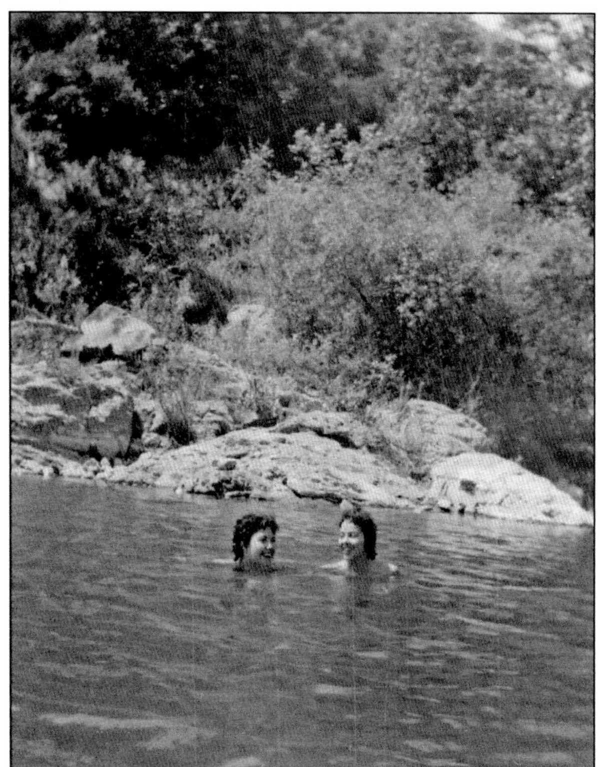

The children enjoyed swimming in Sespe Creek during the hot summers. Summertime fun also included collecting stones and floating along the water on inner tubes. One day, Jerry Crockett disappeared and he was not at any of his usual places. Finally, his parents, Perry and Opal, received word that Jerry had gone across the stream to see his grandmother and his best friend, Kenneth Thompson. The Thompsons lived near Forrest and Mary Wheeler, Opal Crockett's parents. Jerry and Kenneth dug a deep hole and stacked it with wood to build a fire to cook birds they had shot. The only thing was they forgot to make a hole in the cardboard ceiling they made to protect the fire, and they were smoked out. They nearly choked before they got out of there. Kenneth's dad, Homer, could not stop laughing when the boys tried to eat the birds they had cooked.

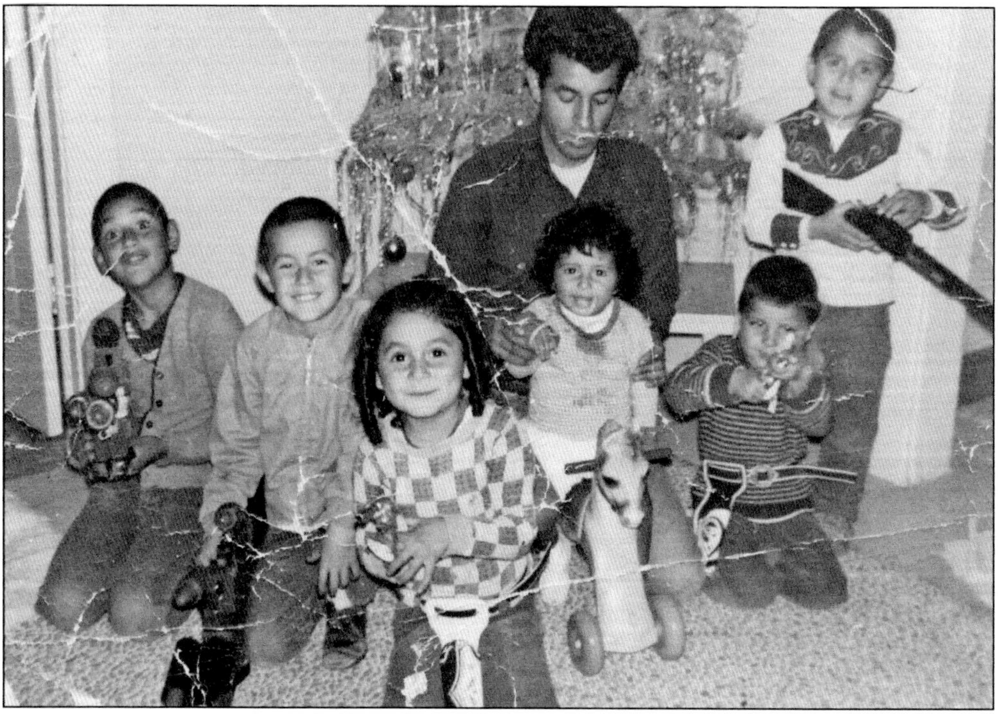

Ernie González is pictured with his sons, from left tor right, Dennis, David, Matthew, Andrew, Mathias, and Ernie.

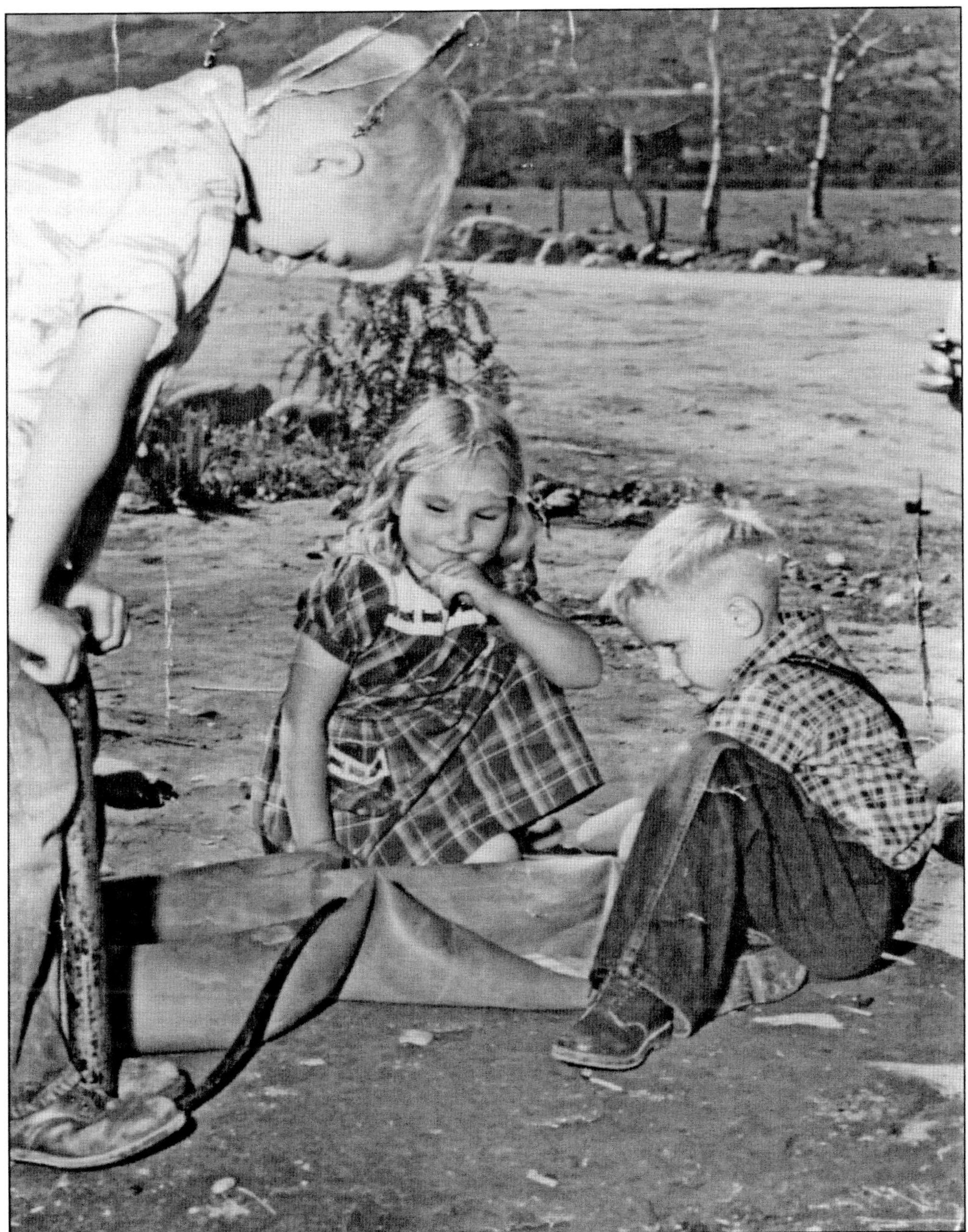

Jerry Crockett (left), Joan Crockett, and cousin DeWayne Wesley Swift are filling an inner tube with air so they can play in the creek. Like many families, the Crocketts moved away from Rancho Sespe for a while, but they moved back. They remained in Rancho Sespe until Oma was a junior in high school. Their home was considered the meeting place, and all the girls would get together and bake cakes, cookies, fudge, and popcorn. During the warm summers, they would all sleep outside. Oma and her friend Barbara Ann Suttle would march down the street wearing their white boots and twirling their batons. Together, they would play kick-the-can and hopscotch, and go roller-skating. They used roller skate keys to tighten their skates around their shoes. (Courtesy of Gail Crockett family collection.)

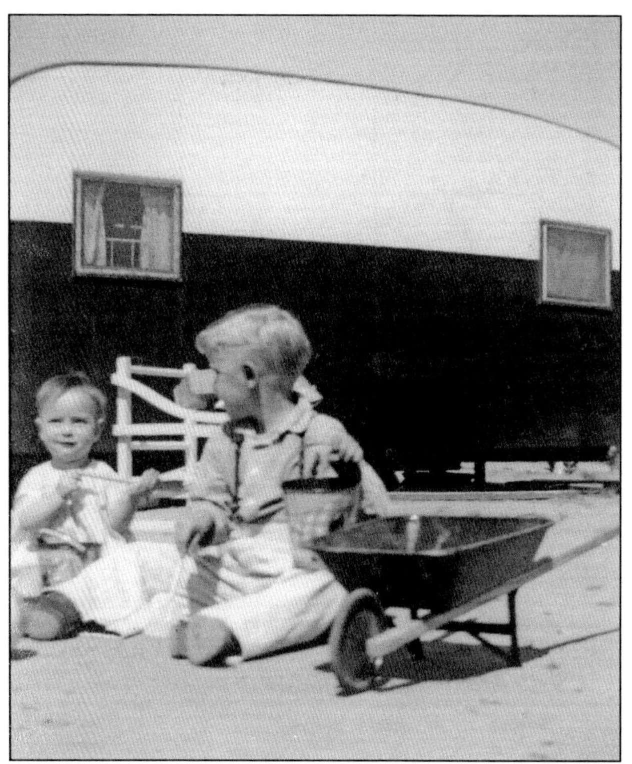

Oma Crockett is shown with DeWayne. Perry Crockett started picking oranges, but the work was too difficult for him, and he was transferred to the bird farm. Turkeys, peacocks, and other birds were housed there. He would take home turkey eggs, and the family enjoyed them as much as chicken eggs. (Courtesy of Gail Crockett family collection.)

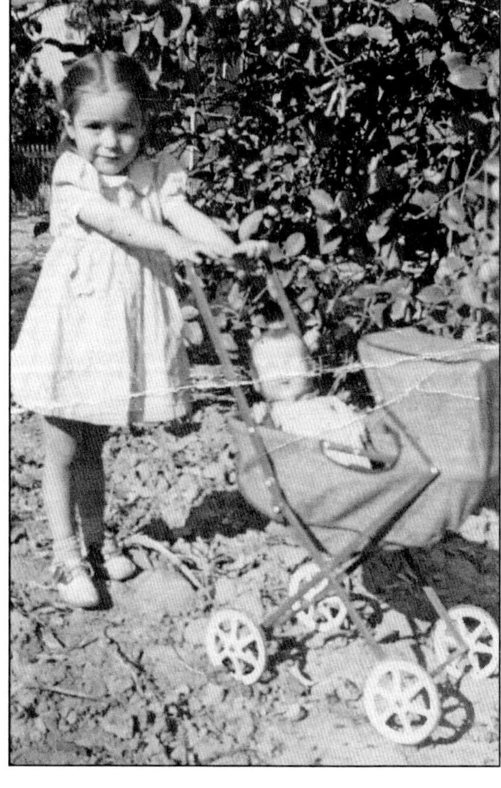

Oma Crockett is pictured with her doll carriage at four years old. The children attended school on the ranch with Martha Stoll as their teacher and Nancy Padelford as her assistant. (Courtesy of Gail Crockett family collection.)

Luther Rodarte is shown here with his brother, Frankie Rodarte. Luther grew up to be a firefighter and worked for the National Park Service. Frankie was Luther's baby brother. Frankie enjoyed tinkering with cars as he grew older. (Courtesy of González family collection.)

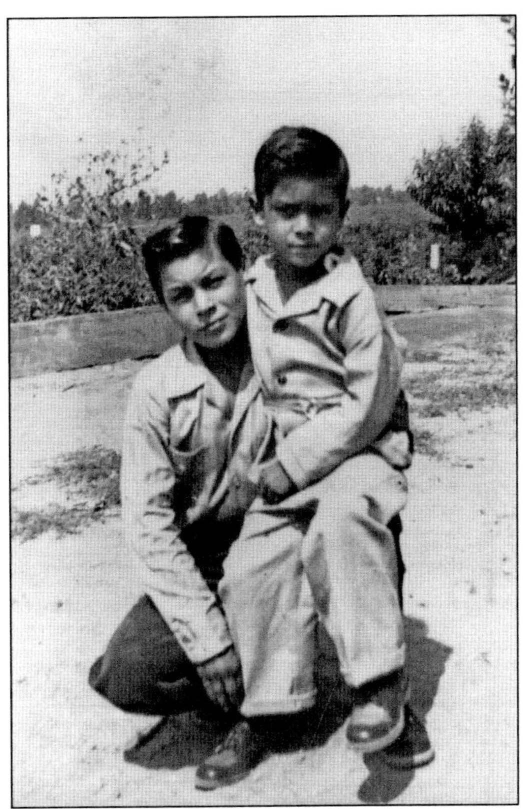

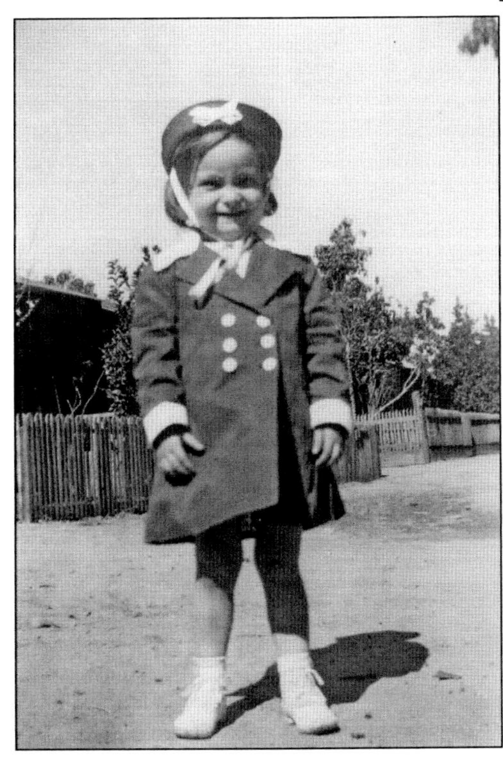

Oma Crockett is pictured in front of her house, No. 84. (Courtesy of Gail Crockett family collection.)

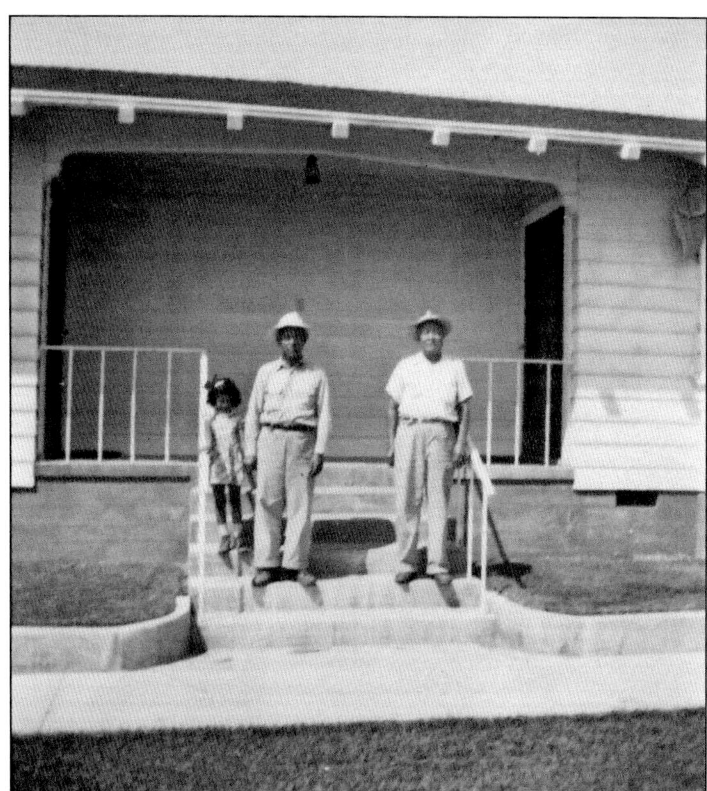

Federico González is working at the ranch school, known as Sespe School. Federico would fix things, as he was a skilled carpenter and handyman. (Courtesy of Larry Tepezano family collection.)

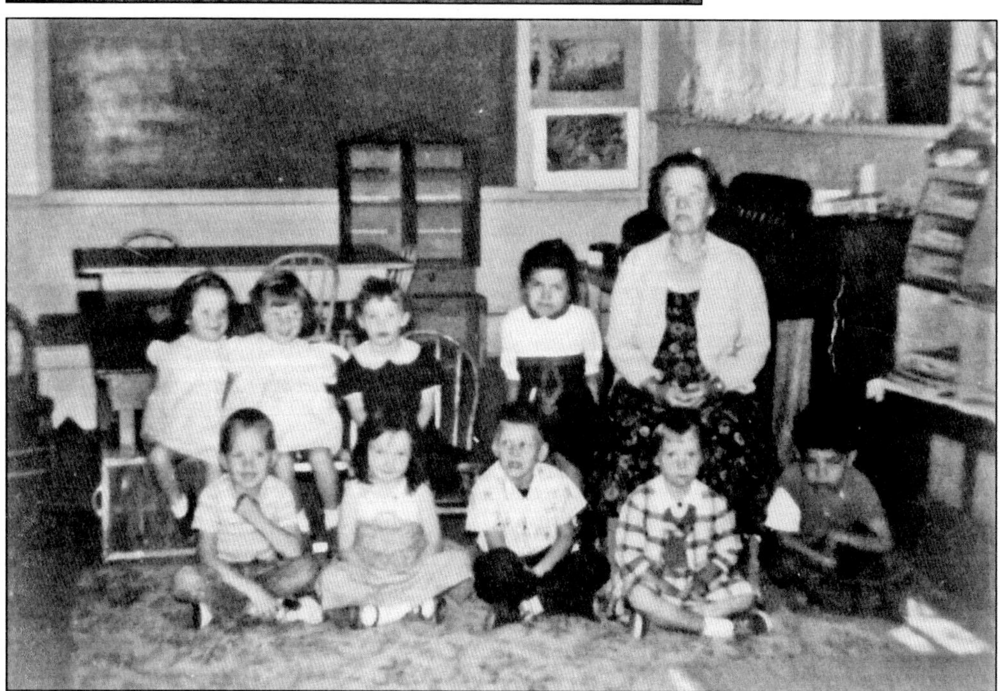

Martha Stoll, the teacher at the Rancho Sespe kindergarten, is shown here with the children. The school was in the center of Sespe Village. (Courtesy of Nancy Padelford family collection.)

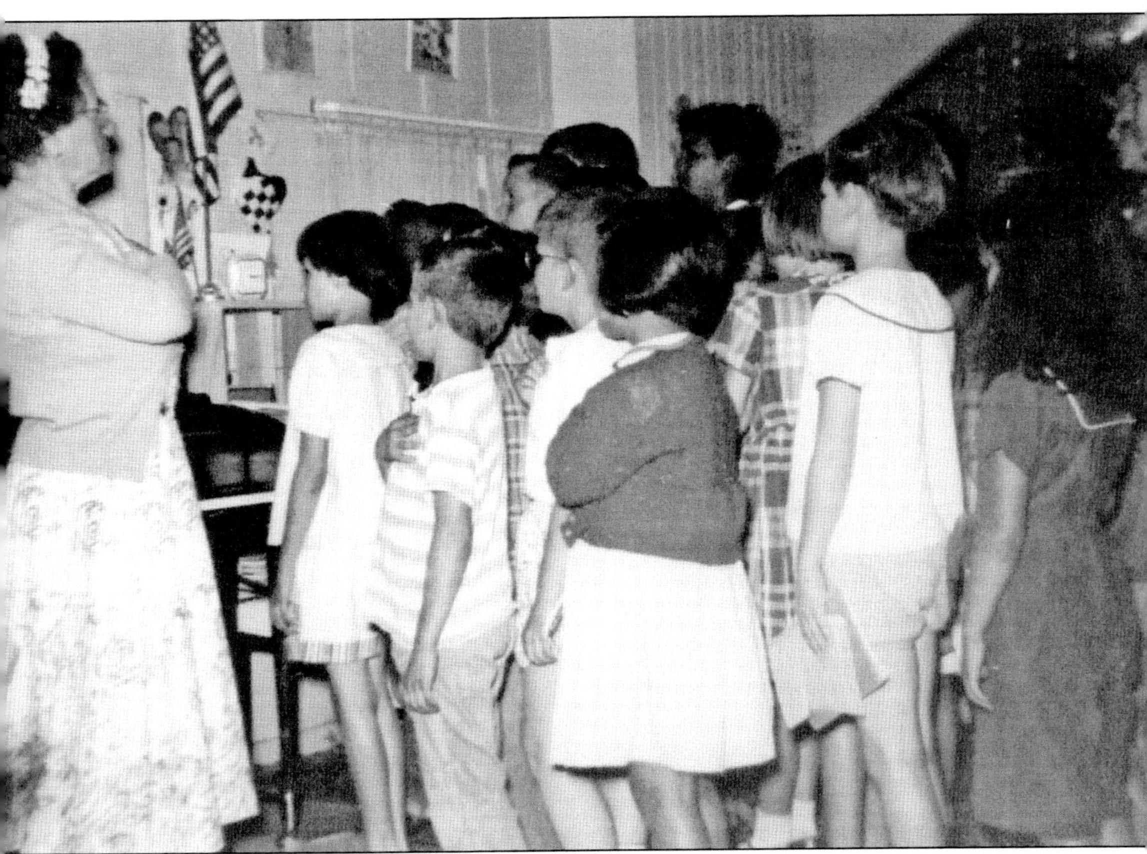

Martha Stoll is leading the flag salute with the children. All the young people spoke highly of her and remembered her as being kind and a very good teacher. (Courtesy of Nancy Padelford family collection.)

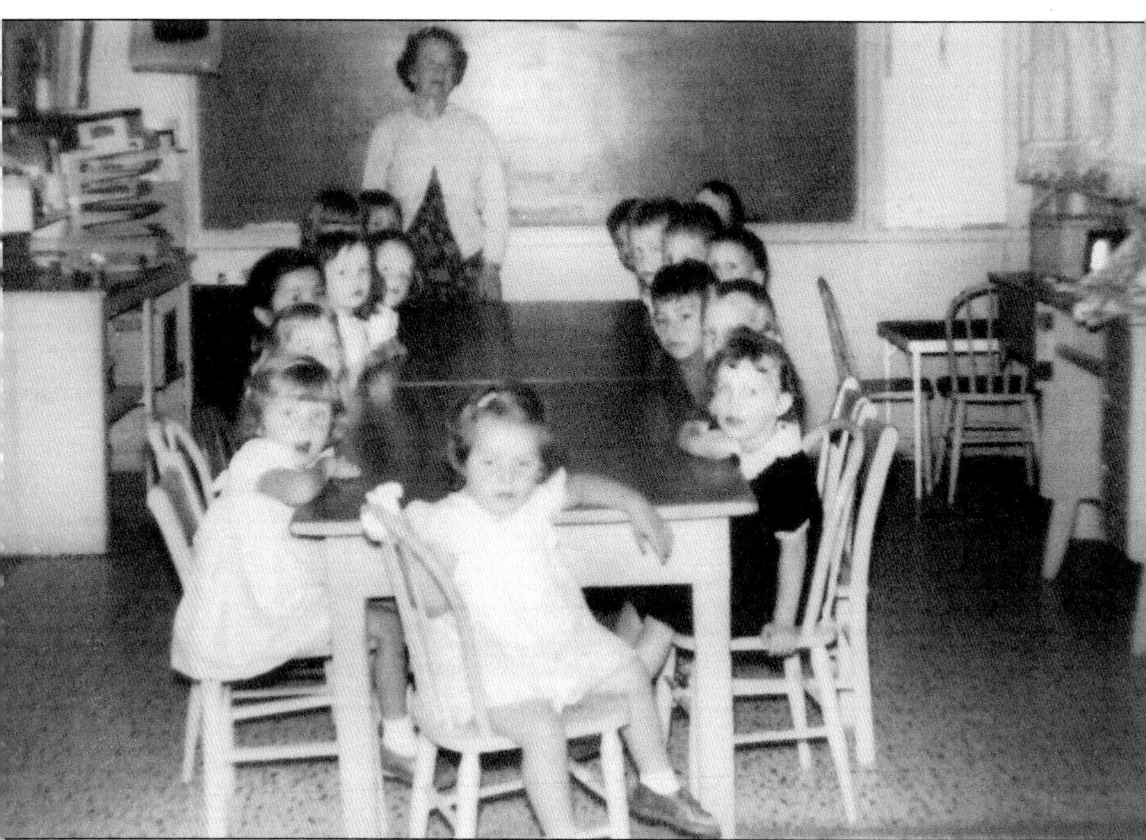

Martha Stoll is pictured with the children in the schoolhouse in Sespe Village. She enjoyed her days as a teacher there. She also helped the girls of the village when they entered Girl Scouts. Becky Sosa Morales recalls when she was a Girl Scout and Stoll helped the girls earn their badges. Henry Ramirez has fond memories about when he attended kindergarten at the Rancho Sespe Schoolhouse. Henry is one of eleven children born to Paul and Guadalupe Ramirez. Henry's brothers are Raul, Paul, Arthur, Tommy, Ramon, and Gilbert. His sisters include Carmen, Esther, Rachel, Linda, and Anita. Henry explains how his sister Esther would walk him to the schoolhouse every day, rain or shine, and the family has good memories of growing up on the ranch property. It was a carefree time then and all were part of a close-knit community. (Courtesy of Nancy Padelford family collection.)

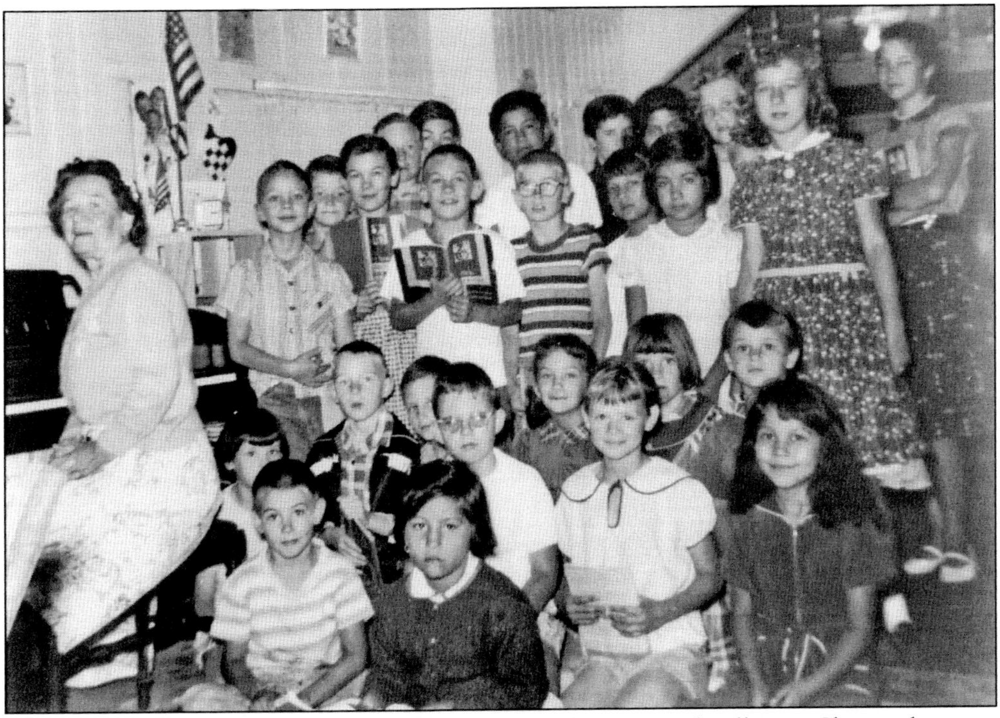

Martha Stoll leads the children in music at the Rancho Sespe schoolhouse. She taught many children to read music and play the piano. (Courtesy of Nancy Padelford family collection.)

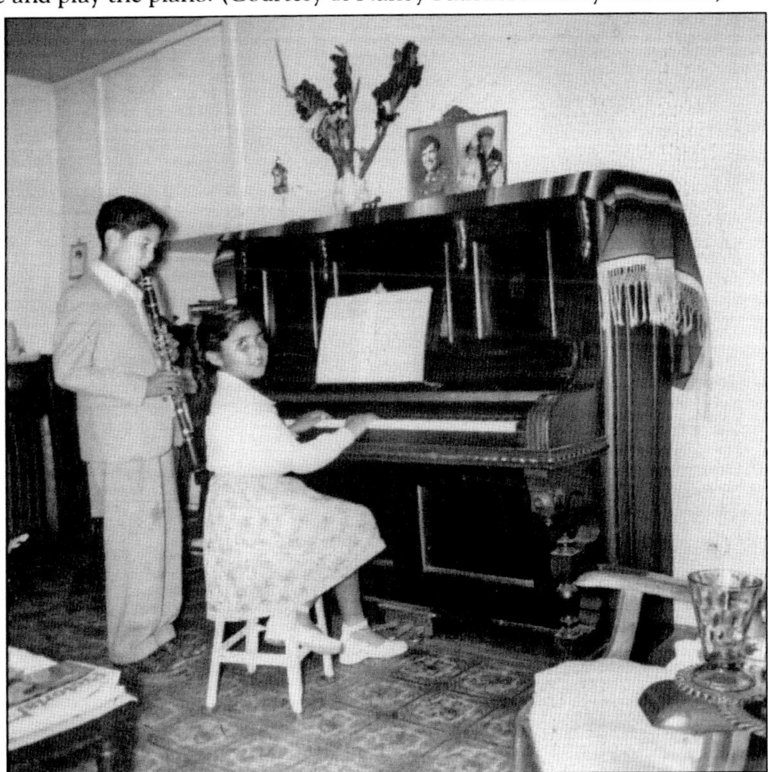

Gilbert Sosa Jr. and Becky practice their music. Gilbert played the clarinet and Becky played the piano. Martha Stoll taught the children who wanted to learn how to play the piano, including Becky.

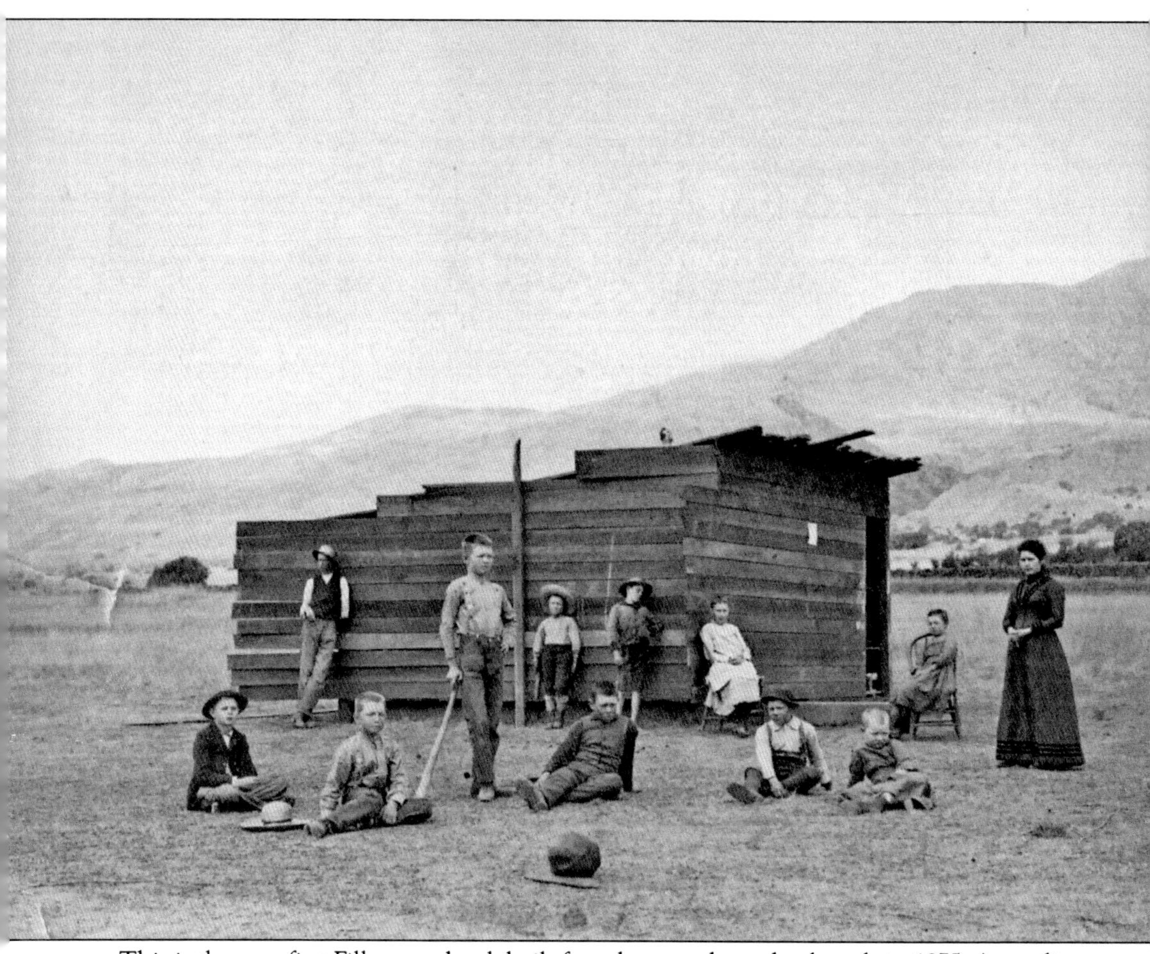

This is the very first Fillmore school, built from borrowed wooden boards in 1875. According to Herman Keene, the boards were labeled with the owner's name, and later, all the boards were returned. Frank Sprague helped to build this first school, known as the Cactus Patch School. Keene explained that while a second schoolhouse was being built, the children attended school temporarily in a tent. (Courtesy of FHM.)

J. D. McNAB, President.
M. DODSWORTH, Vice-President.

F. C. HOWES, Treasurer.
FRANCIS BATES, Secretary.

—OFFICE OF—

SESPE LAND AND WATER COMPANY.

INCORPORATED APRIL, 1886.

ROOM 5, No. 41 SOUTH SPRING STREET,

Los Angeles, Cal., Nov 18 1889

Hon. Board of Trustees
of the Sespe School District
Fillmore Cal.
O. J. Goodenough J. F. McIntyre et al.

Gentlemen

At a meeting of the Board of Directors of the Sespe Land & Water Company, held at the Office of the Company on the 15th inst. a resolution was adopted to donate the following described property for school purposes to wit: Lots #37. 38. 39 40 41 & #42 in Block One (1) in Fillmore Cal. and 130 feet front on Grand Avenue and running easterly two hundred feet. being a portion of farm lot number #149 lying in the South western portion of said lot

Yours Truly
The Sespe Land & Water Co
Francis Bates Secy.

Pictured here is the deed of trust for Sespe School, dated 1889. This was the original Sespe School, located on Sespe Avenue, which became Grand Avenue. (Courtesy of FHM.)

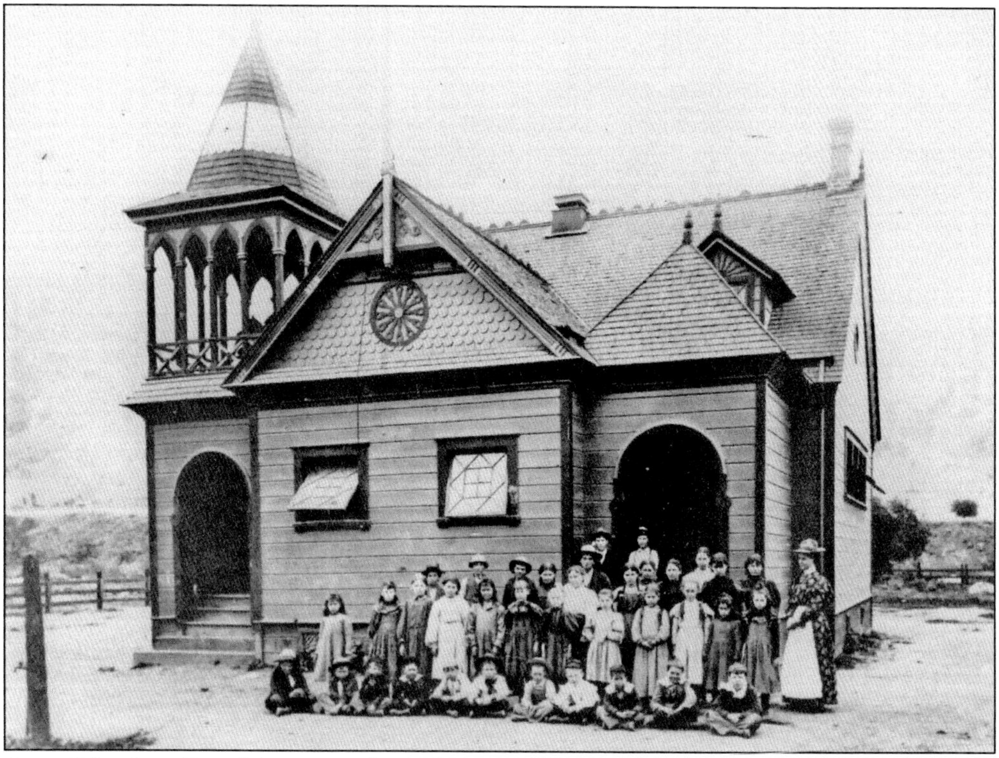

Sespe School on Grand Avenue is shown here in 1894, according to the August 6, 1999, *Ventura Star.* Lily Ann McIntyre is standing a little apart from the rest of the children on the left. Lily remembered going to school barefoot. This school had separate entrances for boys and girls. (Courtesy of FHM.)

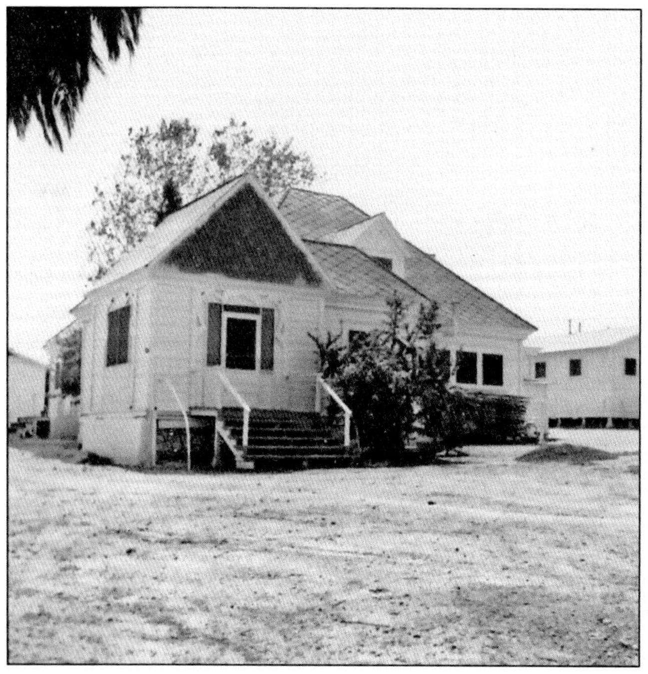

The second schoolhouse was located on Hardison Road north of Highway 126, according to Herman Keene in the 75th-anniversary edition of the *Santa Paula Chronicle*, in 1963. This article explains that the building is still standing and is being used as a Rancho Sespe dormitory and cookhouse. This cookhouse was named a Ventura County Historical Landmark. (Courtesy of FHM.)

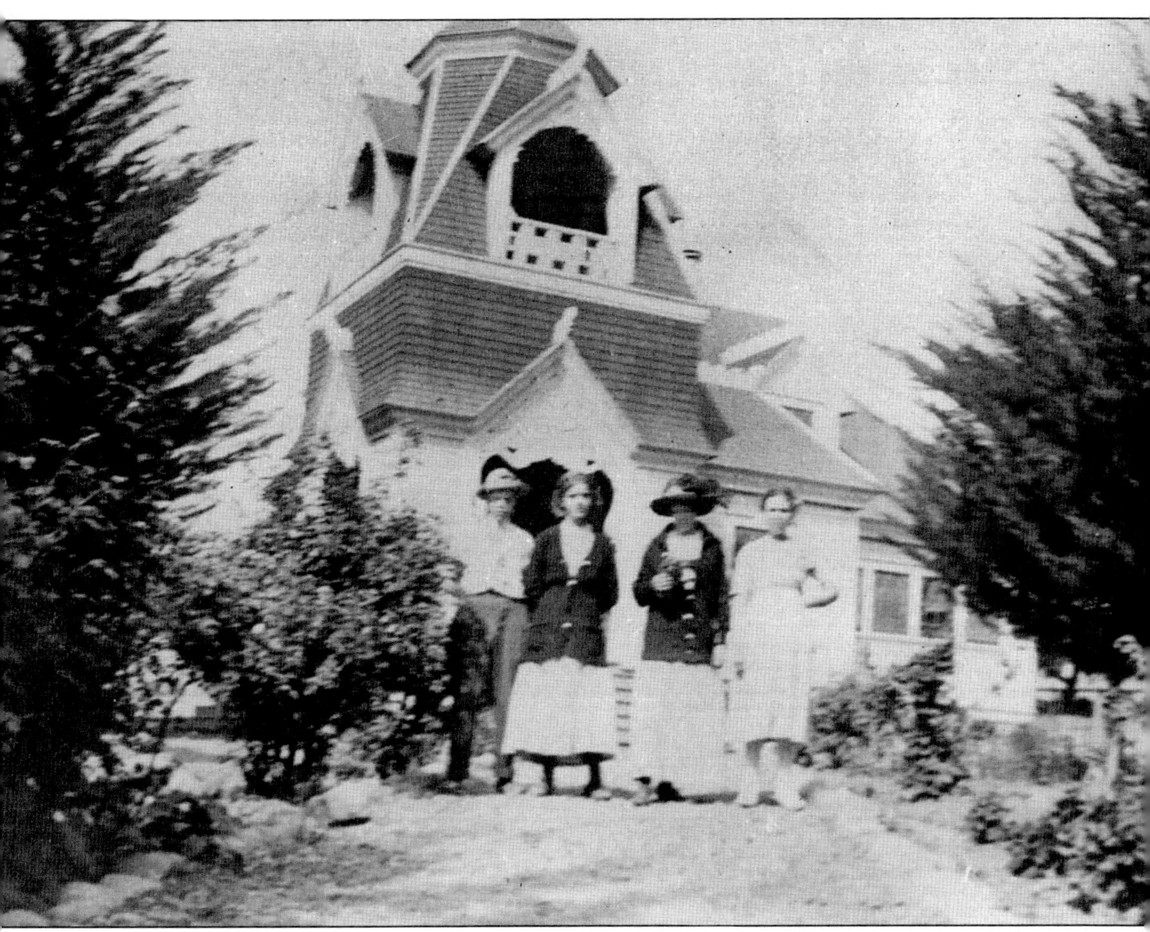

This is San Cayetano School, which was built on Seventh Street in Rancho Sespe. Many Rancho Sespe children attended first through third grades here. Petra González remembers three teachers who taught at this school in 1936, when she was seven years old: a Mrs. Hiller taught first grade; Dorothy Rykeman, second grade; and Gertrude Campbell, third grade. The children enjoyed being bused into town by Clifford Lee to attend fourth to sixth grade. If the children spoke Spanish, they attended Mountain View School. Those who spoke English were transported across town to Sespe School. Henry Ramirez recalls his experience on the Mountain View playground as a difficult one because there was no grass, only dirt. The playground was covered with rocks of all sizes. Henry recalls how the basketball court at Mountain View School had nails sticking out in some places. Hazel Hiberly was principal at Mountain View School, and she became principal of the new San Cayetano School when it opened in 1957–1958. (Courtesy of FHM.)

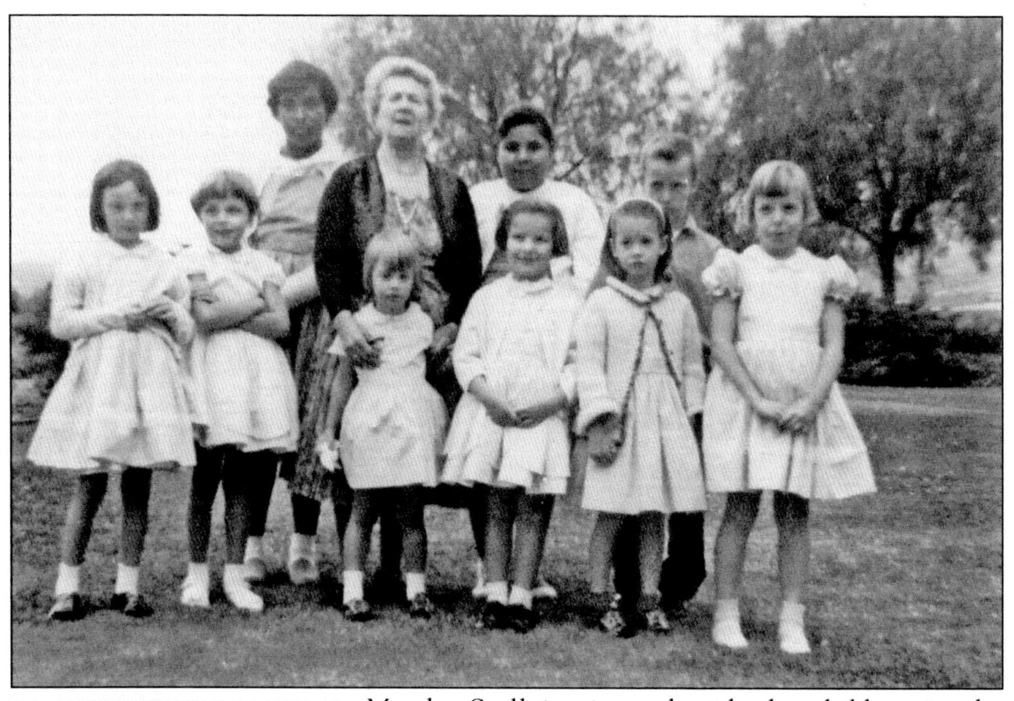

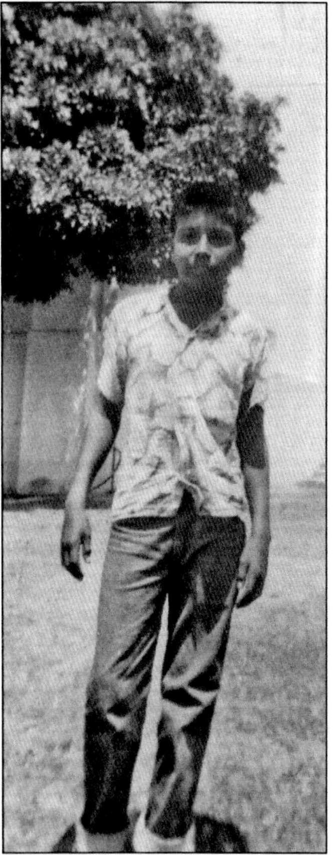

Martha Stoll is pictured with the children in the schoolyard in Sespe Village. (Courtesy of Nancy Padelford family collection.)

Tommy Ramirez is standing in front of Fillmore High School at approximately 12 or 13. He grew up and had six beautiful children of his own. He and his former wife, Mary, emphasized education for their children. Their eldest son, Vincent, is a registered nurse in the surgery department of Community Memorial Hospital in Ventura. John is a California highway patrolman, and Larry is a licensed X-ray technician. Another son, Richard, lives with his wife, Donna Davies, in Sacramento, where Dr. Ramirez serves as a counselor and professor. James is also in education, as are Tommy's two daughters, Carol and Tiffany. Tommy and Henry have fond memories of growing up in Rancho Sespe. It was common for boys to seek a work permit at the age of 13. Henry worked during the summers pulling weeds on the ranch. He also painted the trunks of lemon trees and sprayed weeds alongside Juan Rodarte; helped pick lemons, oranges, and grapefruit (when he learned how difficult that was, he aspired not to do that type of work for the rest of his life); and helped the Thompson brothers graze some of the cattle on the ranch. When the Ramirezes lived in house No. 23 in Sespe Village, Henry remembers all the parades for the Cinco de Mayo and September 16. These celebrations united the Rancho Sespe community. After a while, the Ramirez family moved into house No. 128, where the Ramirez boys made new friends. (Courtesy of Ramirez family collection.)

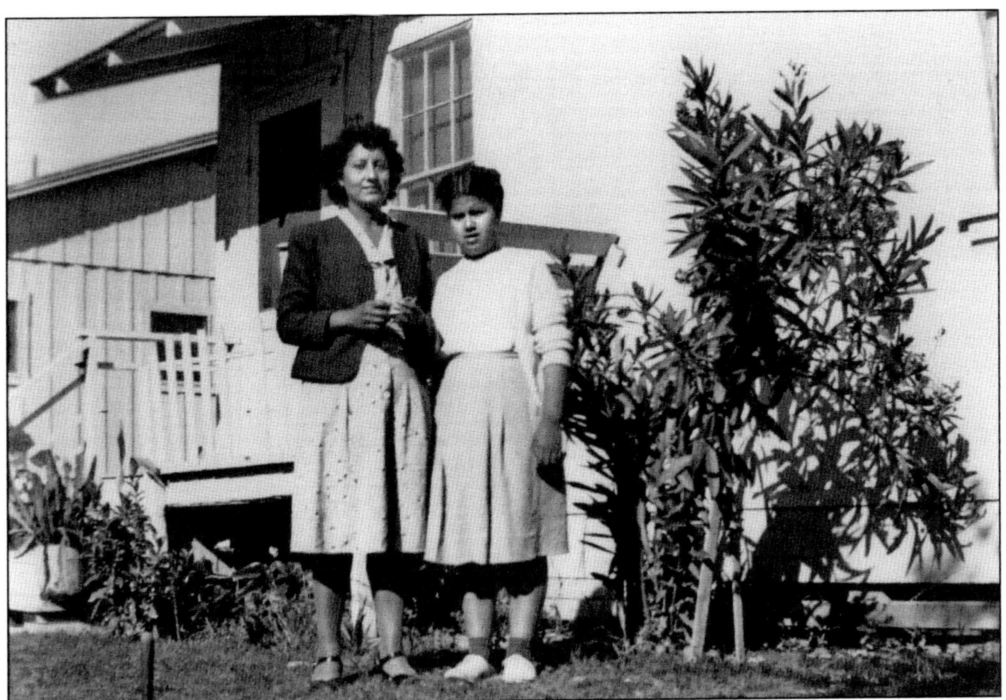

Fannie Sigala is shown here with her daughter, Betty Sigala. Fannie is one of Becky Sosa's aunts.

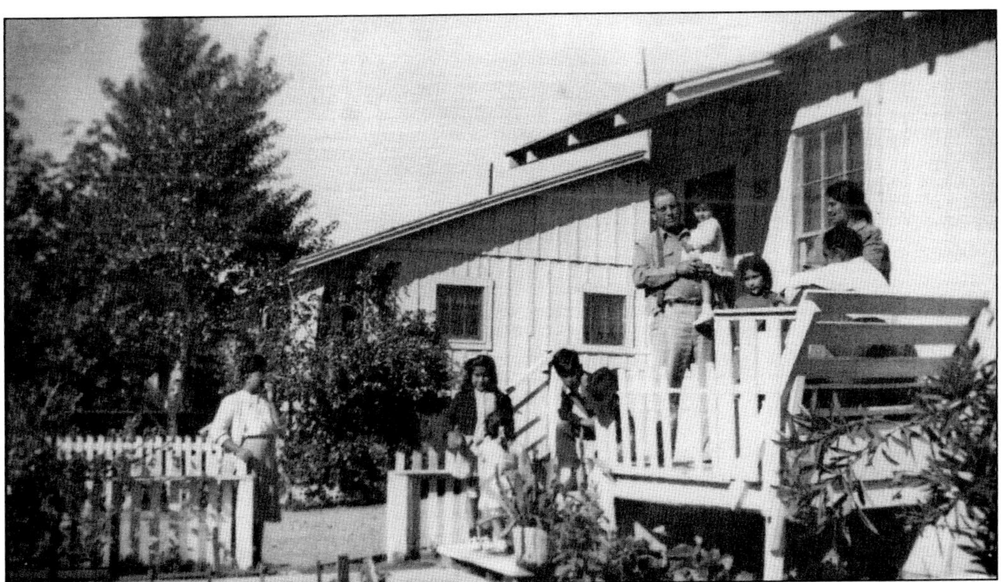

At a Sigala family gathering are, from left to right, Betty Sigala, Helen Sigala, Gilbert "Mo" Sosa, Tony Sigala holding Rosemary Sigala, Becky Sosa, Mary Valenzuela, and Tom Valenzuela. The family would gather to celebrate holidays, birthdays, Thanksgiving, and Christmas with dinners and by making the traditional tamales.

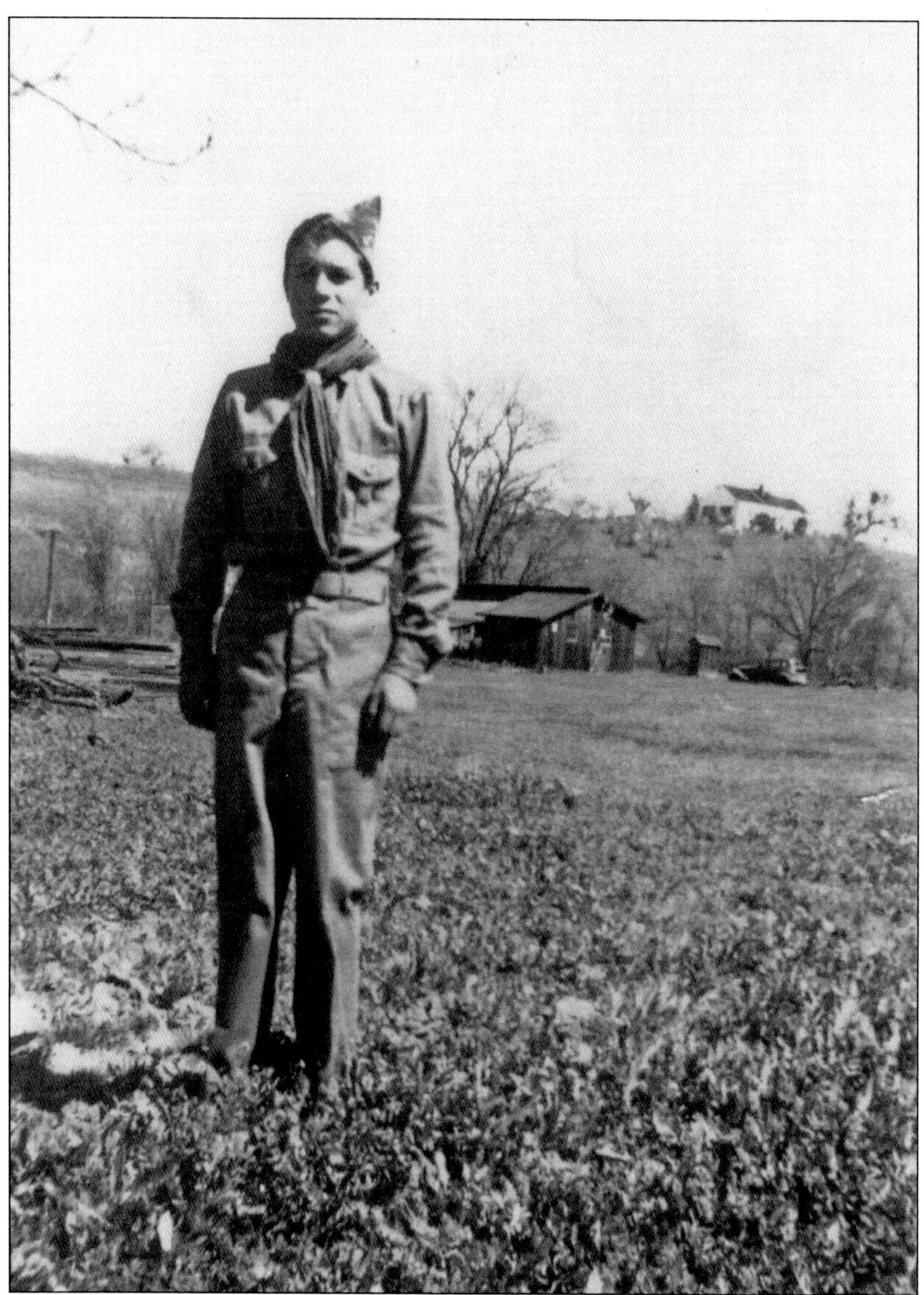

Luther Rodarte is shown in his Boy Scout uniform at Rancho Sespe. The community had a Boy Scout troop, and Luther was very active in that organization. Scoutmaster Dave Diffenduffer took the boys camping and swimming. Those who belonged to Troop 405 recall what a wonderful time it was for them. Sadly, Diffenduffer perished during a solo flight a few years later. The boys of Rancho Sespe lost a role model and friend.

John and Betty González are shown here. They were part of the Federico González family, a very close-knit family that worked on the ranch. (Courtesy of Larry Tepezano family collection.)

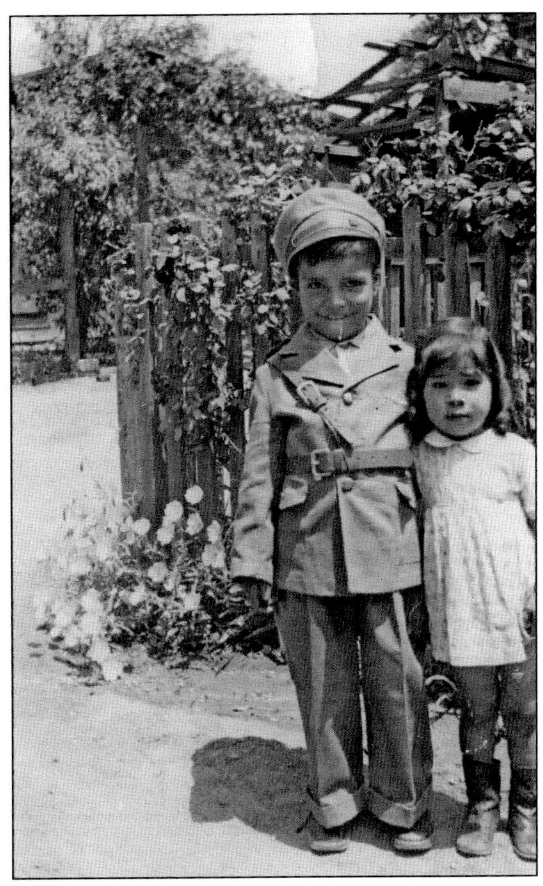

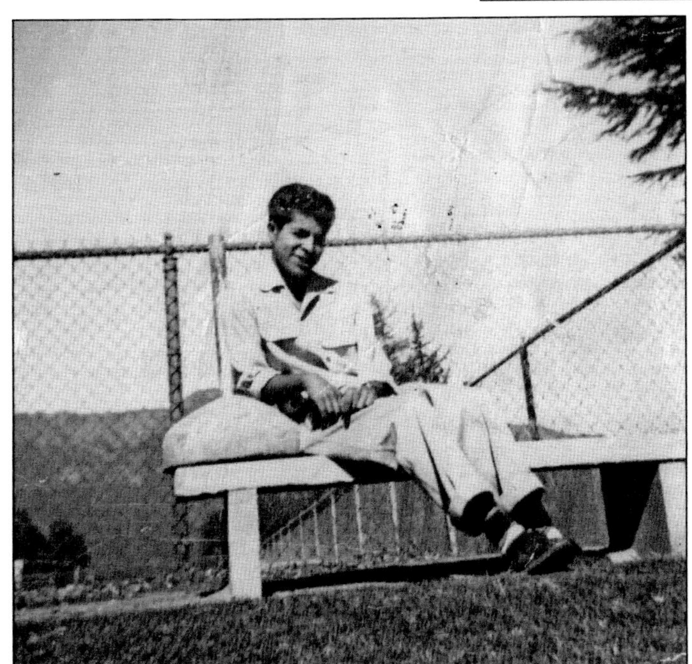

George "Rudy" González is pictured at 16. He joined the Army at 19. Later, he married, and he and his wife had three children. Rudy returned to work at the packinghouses in Santa Paula, Ventura, and Fillmore, then he worked at the Fillmore Insectary for 20 years. (Courtesy of Larry Tepezano family collection.)

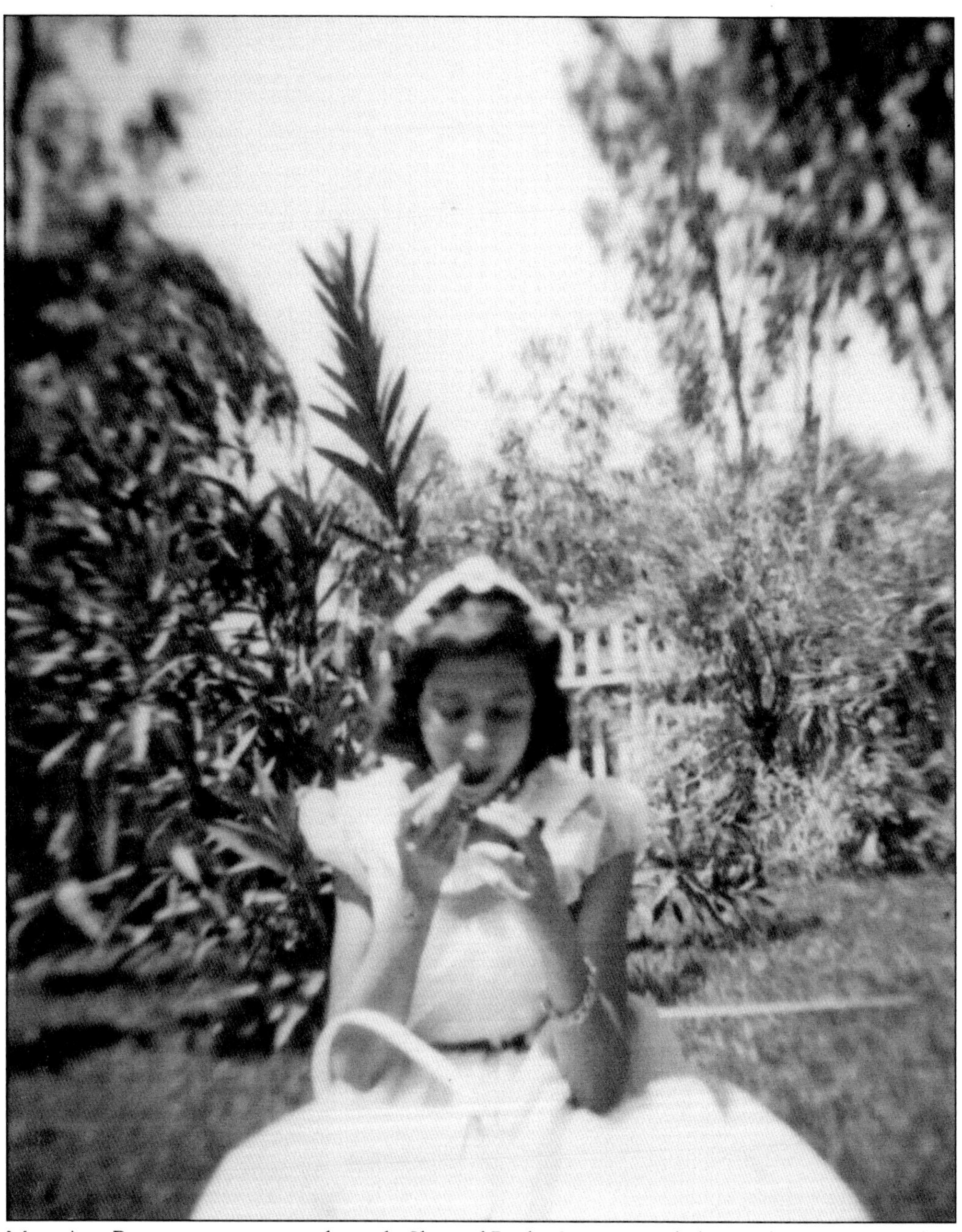

Mary Ann Renteria is putting on lipstick. She and Becky Sosa enjoyed playing dress-up, pretending to be *comadres*, and playing with their dolls. Mary Ann and Becky would talk about their make-believe children and walk around with their dolls as they pretended to go shopping. Becky remembers Kenneth "Speed" Stewart dressing up as Santa Claus one year and delivering a doll and a new baby carriage to Becky for Christmas. She loves that doll and has kept it all these years.

Margie and Larry Tepezano are shown at the bungalow, playing by the orchard. The Spalding main house, the guesthouse bungalow, and the wall that runs along the private drive to the Spalding main house are designated historic landmarks. (Courtesy of Larry Tepezano family collection.)

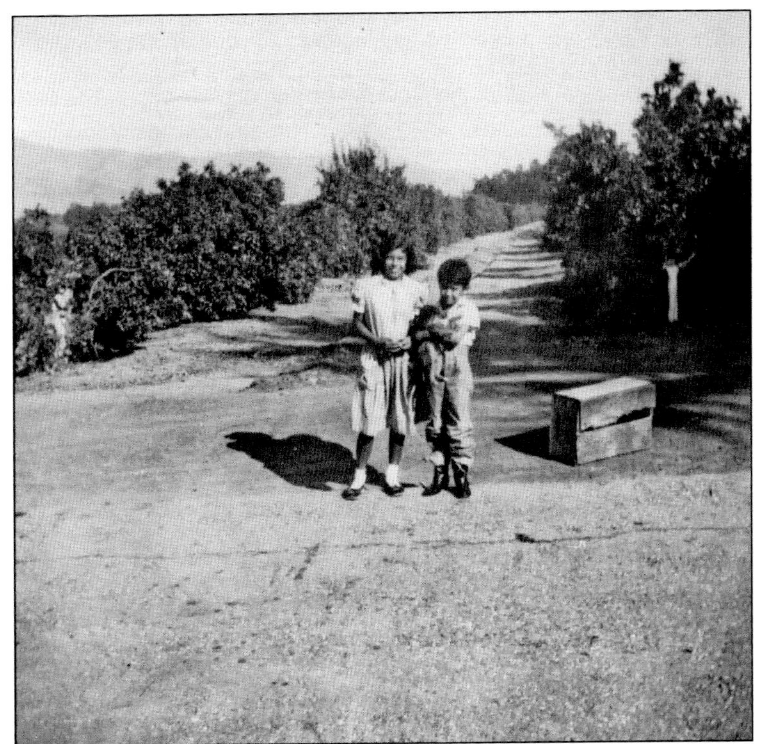

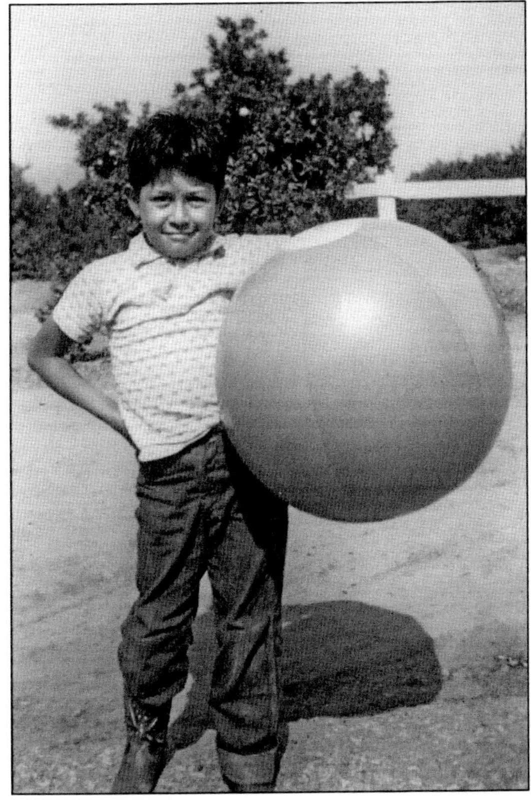

Larry Tepezano is shown here playing with a ball on the Spalding grounds. Larry's mother, Connie Tepezano, was a cook for the Spaldings. (Courtesy of Larry Tepezano family collection.)

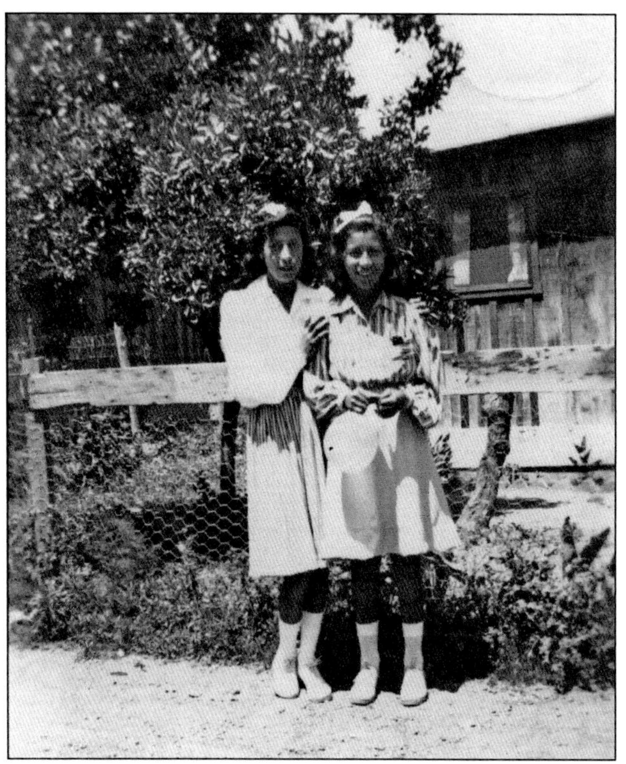

Connie and Mary González are pictured here. Connie González Tepezano worked for the Spaldings. Mary later married, and she and her husband, Manuel Santiago, lived in Oxnard with their three children. (Courtesy of Larry Tepezano family collection.)

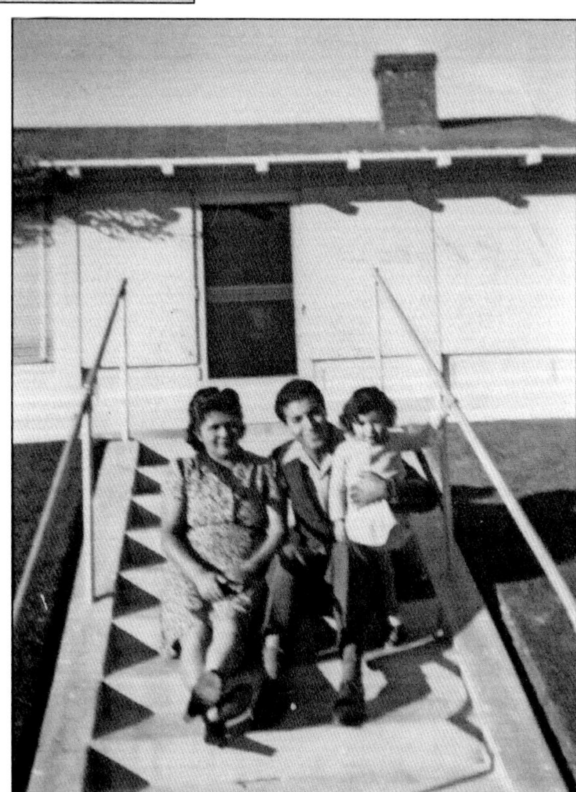

Members of the González family pose in 1946. The close-knit family, with five sons and four daughters, celebrated all the holidays together and worked hard.

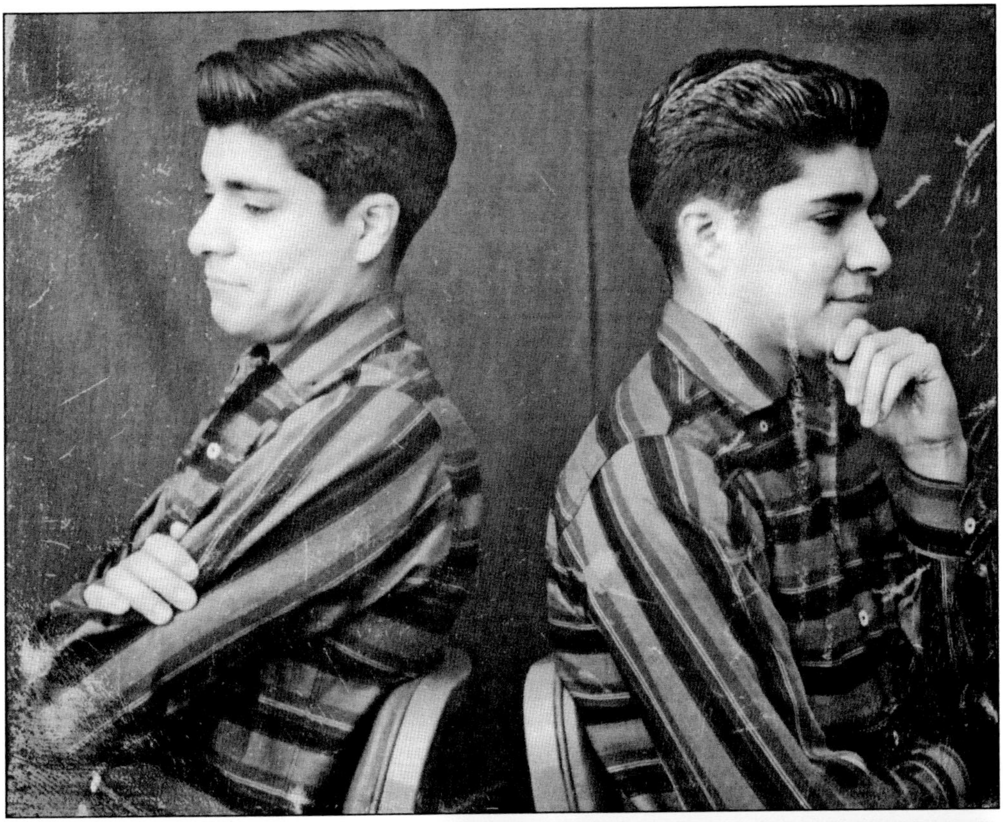

John "Shorty" González is shown in this double exposure. He wrote on the back, "It's bad enough with just one . . ." John married and had three children, then he retired from the Air Force after serving for 26 years.

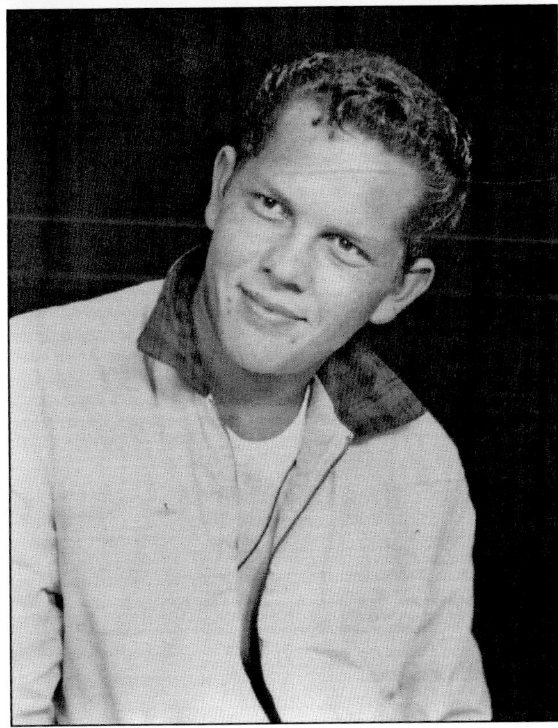

Paul Rueda (pictured) and John "Shorty" González were the best of friends.

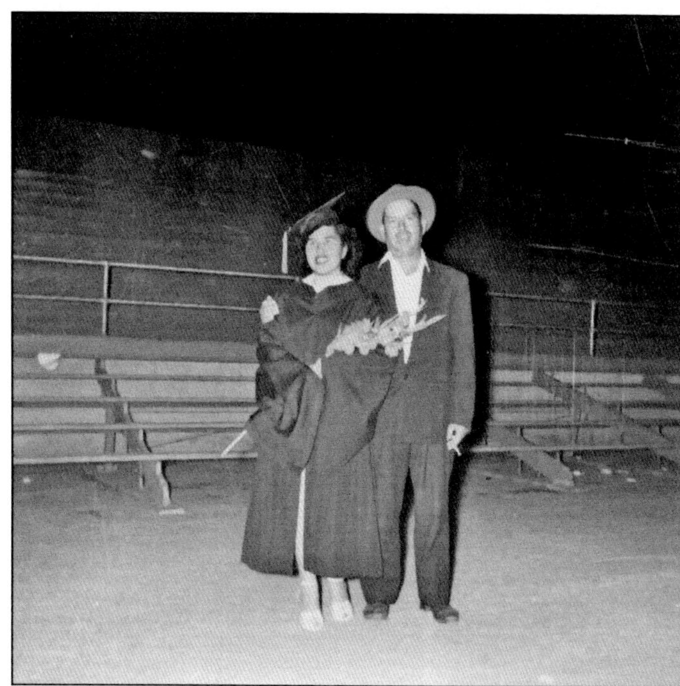

This photograph was made at Glafida Sigala's graduation from Fillmore High School in 1949. She is pictured with her father, Panfilo Sigala.

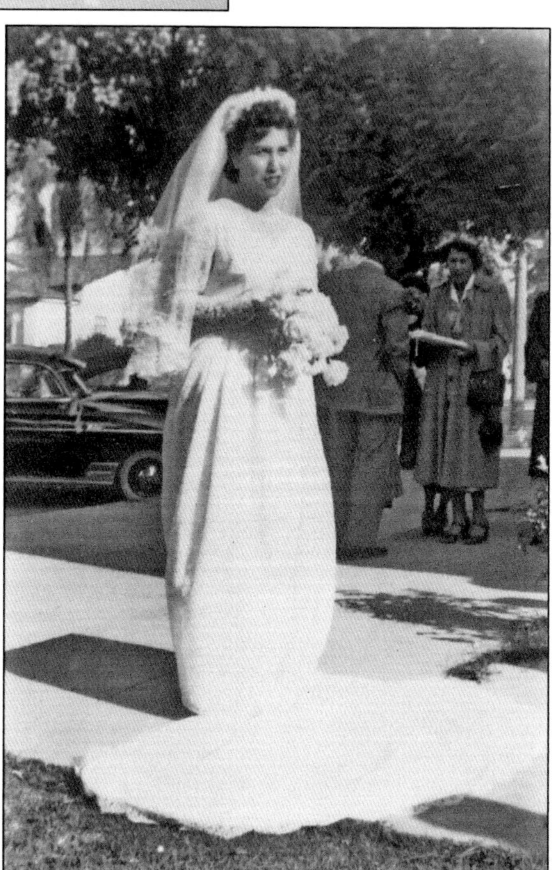

Glafida Sigala is pictured on her wedding day in August 1953. "Fida" and Luther P. Rodarte were married at St. Francis of Assisi Catholic Church in Fillmore. Her dress was made by her sister, Alice Sosa.

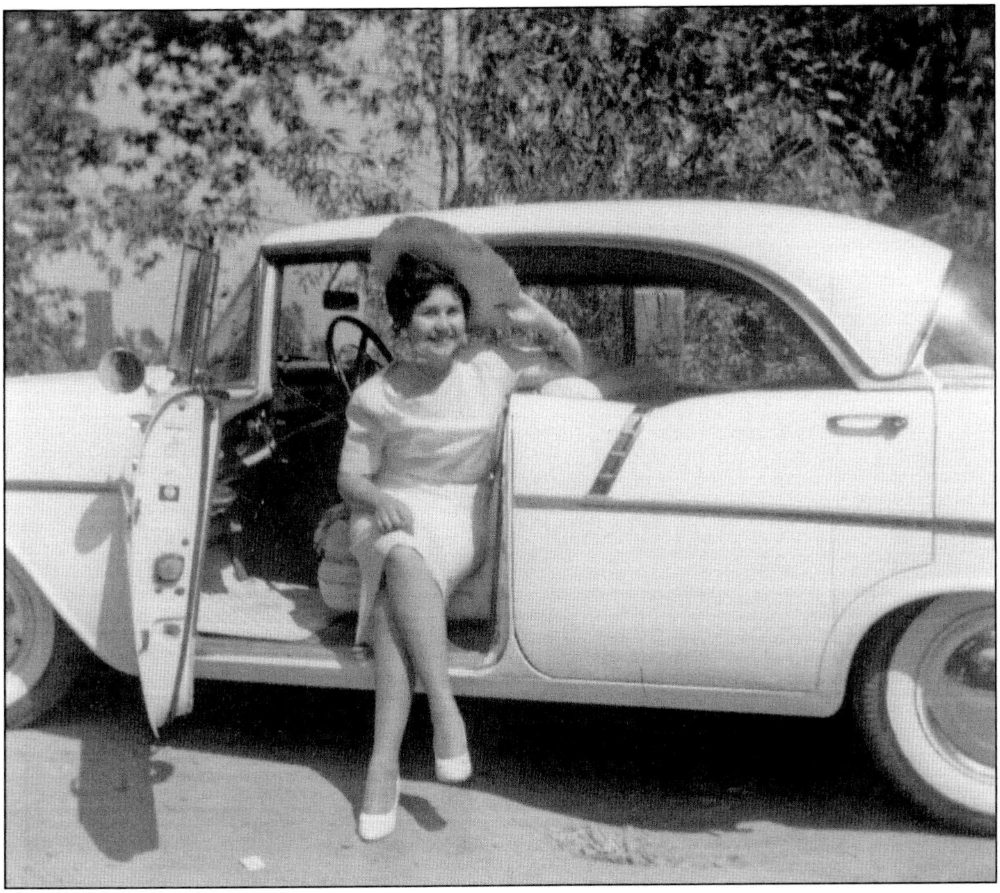

Becky Sosa sits in Ernie Morales's new white 1956 Chevrolet. She remembers taking daytime road trips in that car.

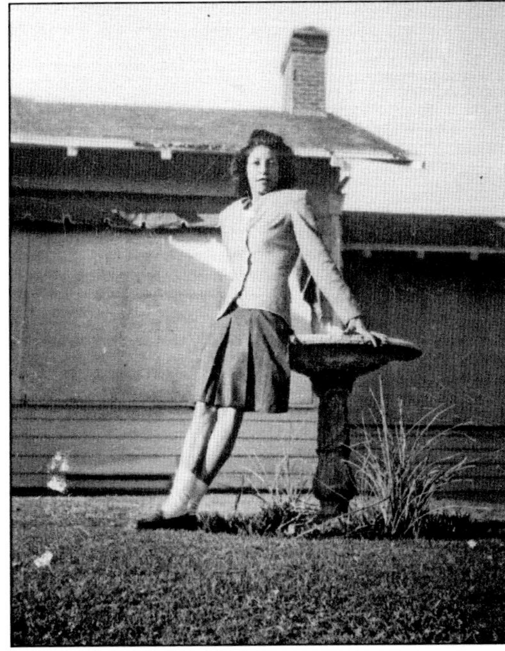

Connie Tepezano is pictured at the bungalow. She lived at the Spalding guesthouse with her husband, Larry, and their children.

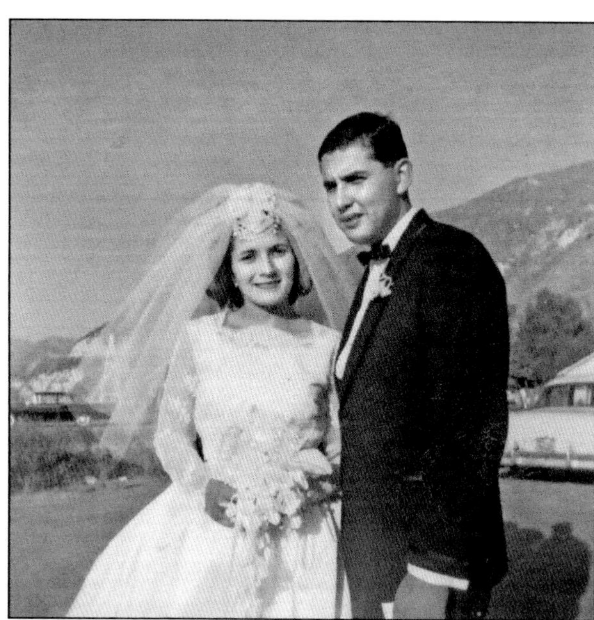

Gilbert Sosa Jr. and Betty Frias were married on December 1, 1962. Their wedding day is pictured.

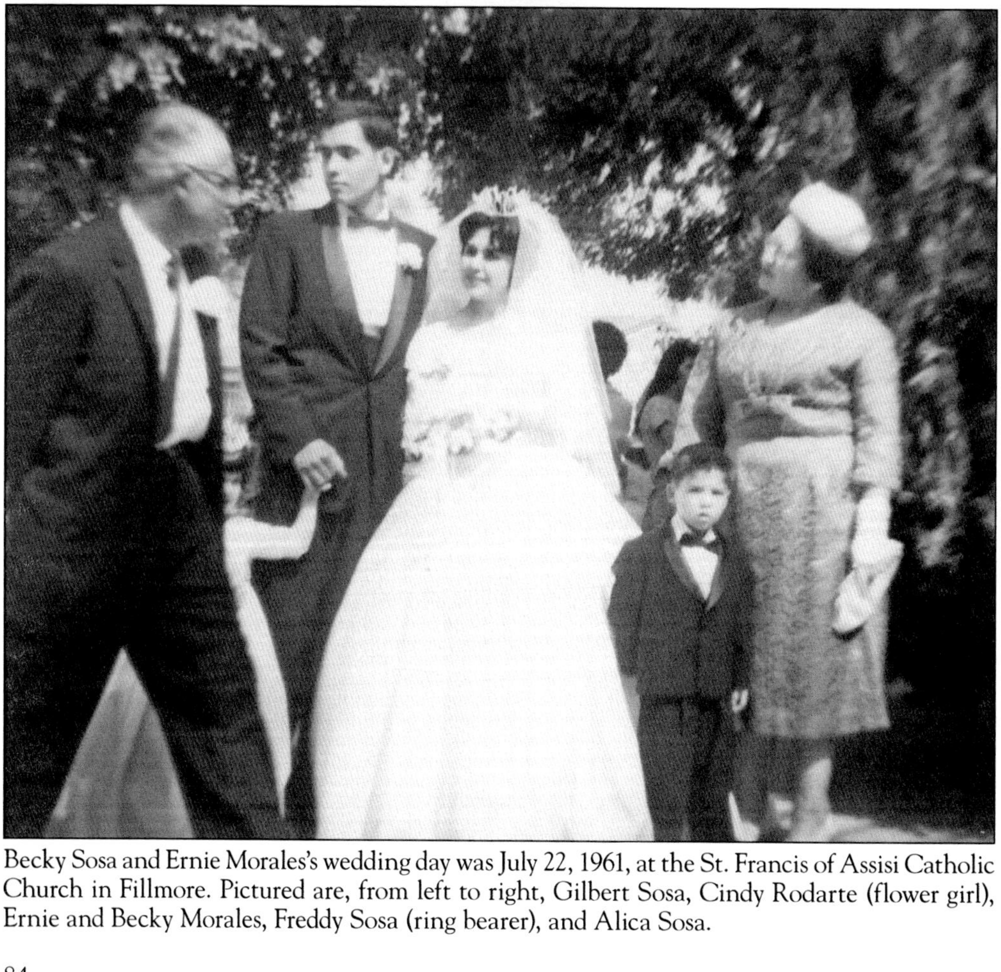

Becky Sosa and Ernie Morales's wedding day was July 22, 1961, at the St. Francis of Assisi Catholic Church in Fillmore. Pictured are, from left to right, Gilbert Sosa, Cindy Rodarte (flower girl), Ernie and Becky Morales, Freddy Sosa (ring bearer), and Alica Sosa.

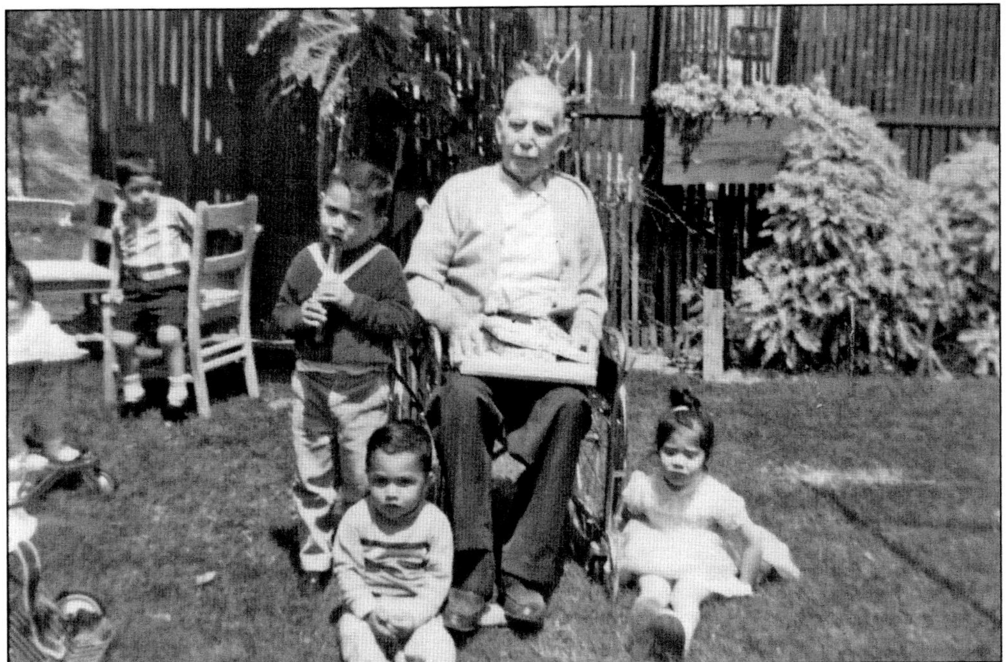

The Sosa grandchildren are pictured here with their great-grandfather Fernando Sosa. Jeff Morales is sitting on a chair, Ray Sosa is next to great-grandpa Fernando, Steve Sosa is sitting on the lawn with Gina Sosa, and Brian Morales is in the stroller (partially obscured at left).

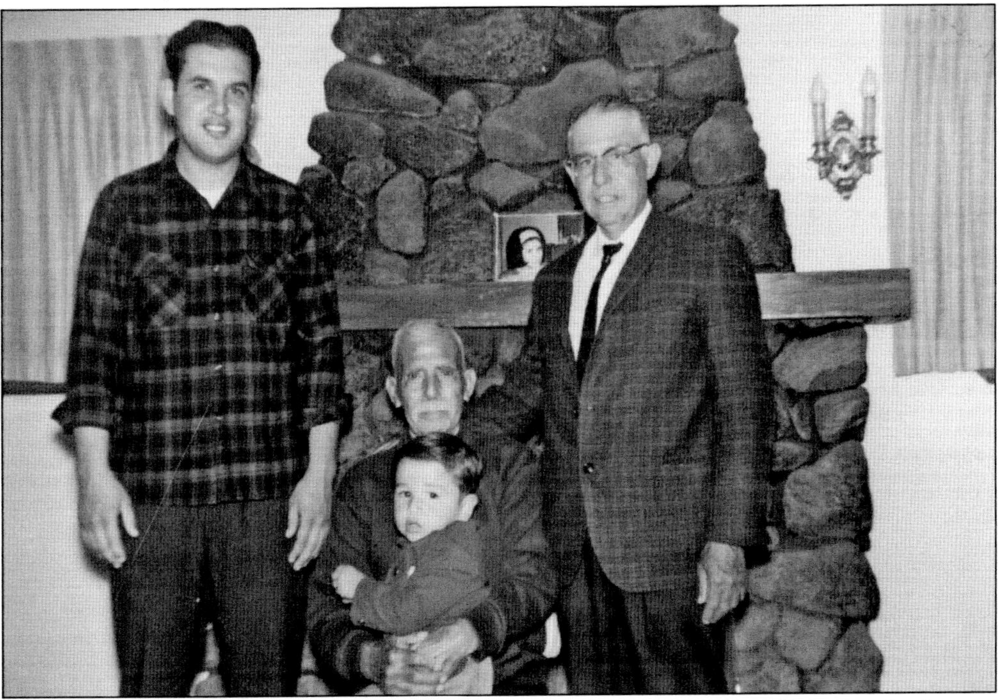

Four generations of Sosas lived in Rancho Sespe. From left to right are Gilbert Sosa Jr., his son Ray Anthony Sosa, great-grandfather Fernando Sosa, and grandfather Gilbert Sosa Sr. All lived in Rancho Sespe.

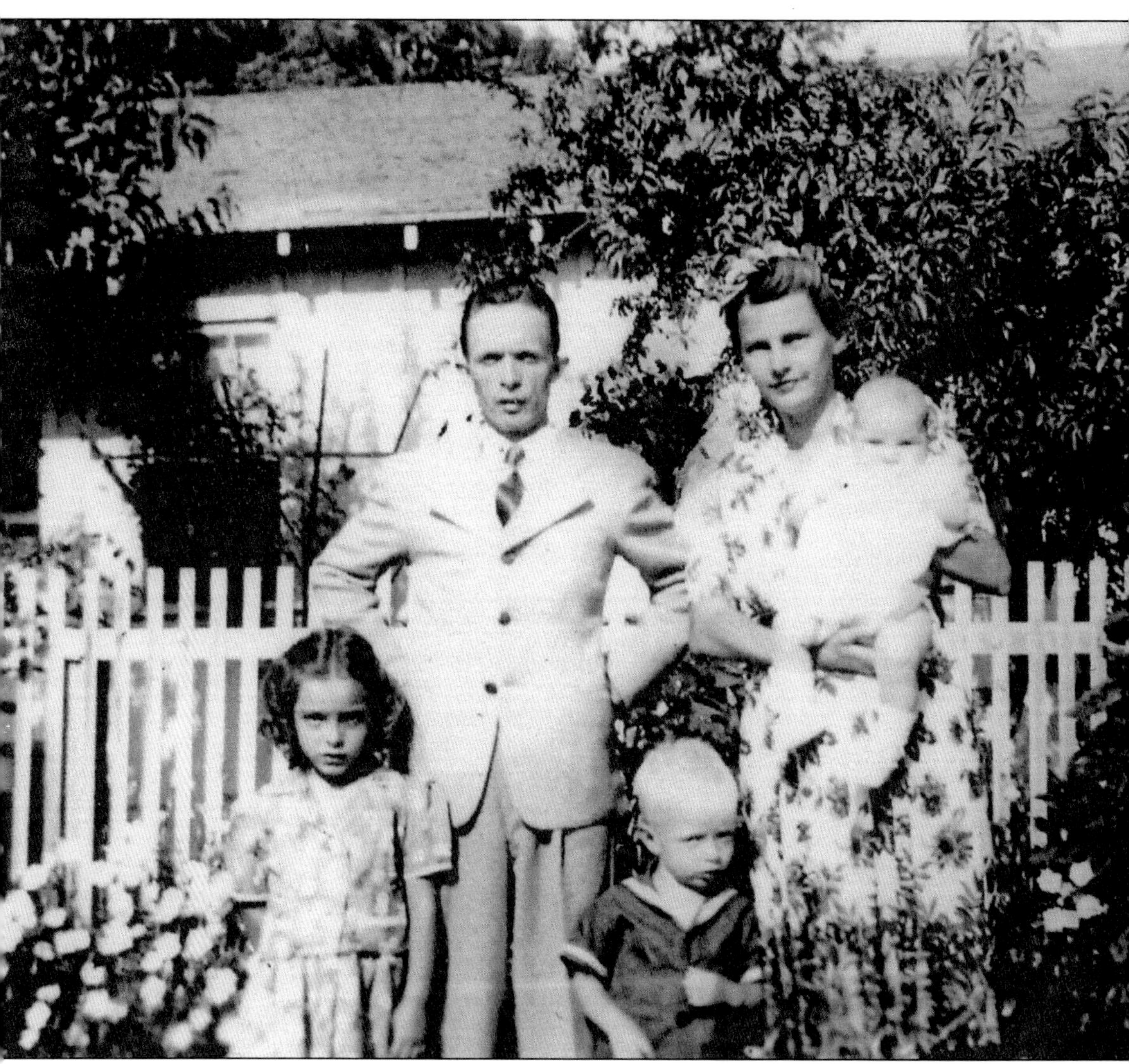

Perry Crockett poses with his three children in 1944. Opal Wheeler Crockett is holding her daughter Joan. Oma Crockett and her brother, Jerry Crockett, are featured at the bottom of the photograph. They lived in house No. 84 in Rancho Sespe. Families bought their goods at the local store, which Albert Reyes managed. The store provided some groceries and various household items. (Courtesy of Gail Crockett family collection.)

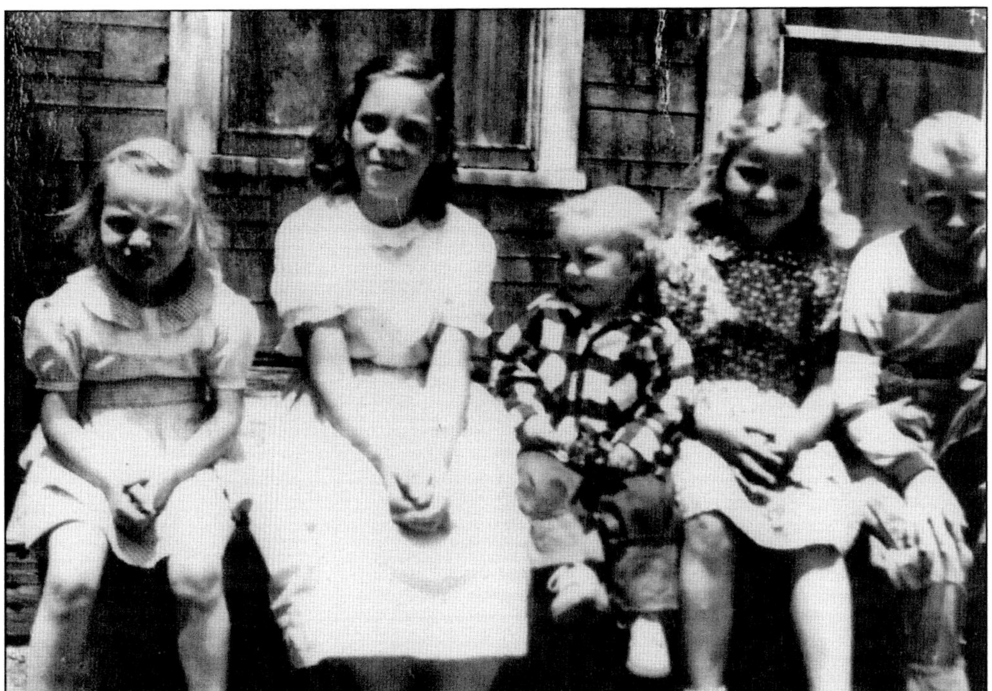

The Crockett children are shown here in 1951. They are, from left to right, Gail, Oma, Dale, Joan, and Jerry. (Courtesy of Gail Crockett family collection.)

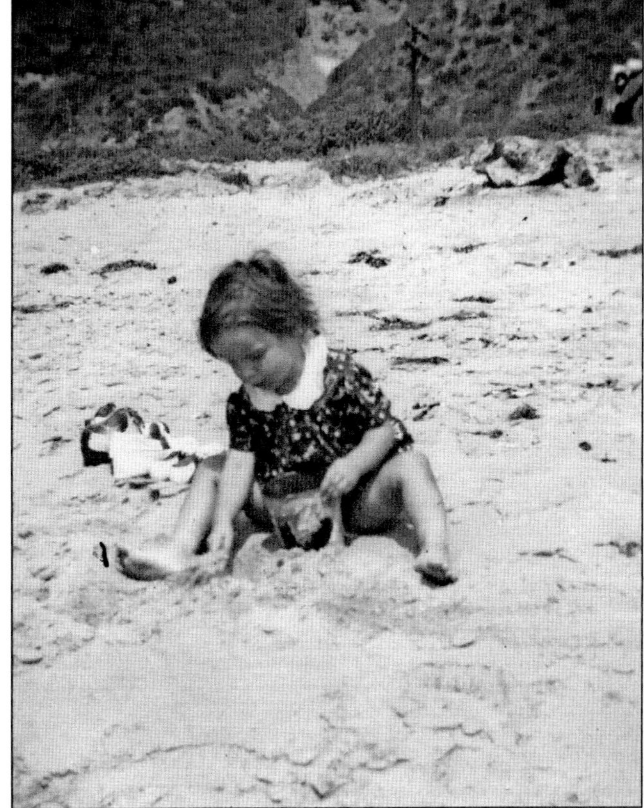

Oma Crockett Burgett is pictured here at the beach. As often as families could, they would drive to the beach in neighboring Ventura. The summers were a time when families could take a picnic lunch and lounge on the sandy shore while the children played in the water. (Courtesy of Gail Crockett family collection.)

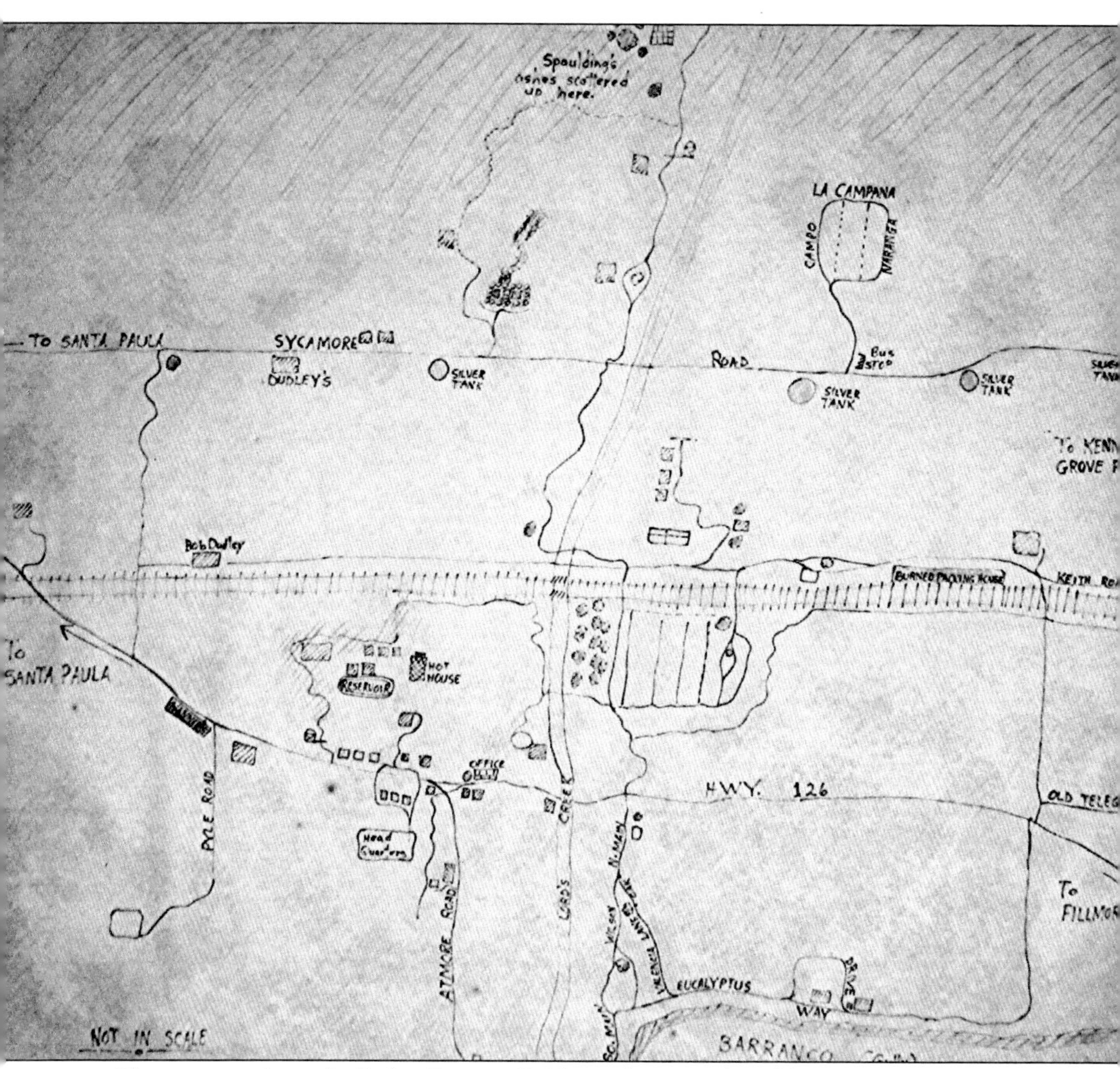

This map was drawn by Esther Ramirez Robles as she remembered how Rancho Sespe looked to her while she lived there with her family. (Courtesy of Ramirez family collection.)

Five

HEROES

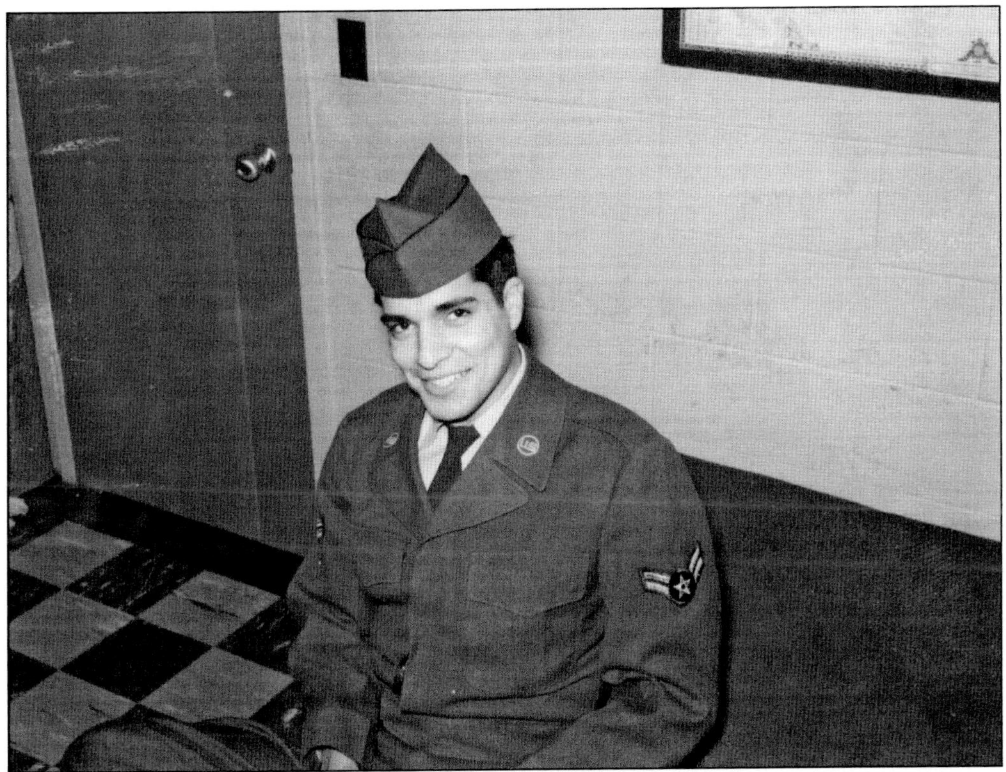

John "Shorty" González is in his Air Force uniform. When World War II broke out, the men enlisted and left to fight for the United States. Women went to work in place of the men, and children who were 14 or older could also work to help the war effort. The ranch had its own packinghouse and was self-sustaining, so women who lived at the ranch could work there. Some of the girls who were old enough could also work. Petra González was one of those who worked at the packinghouse, but she and the other girls could only work for two hours after school due to the child labor laws. Pay in the beginning was 40¢ per hour. During the summer, they were allowed to work more hours.

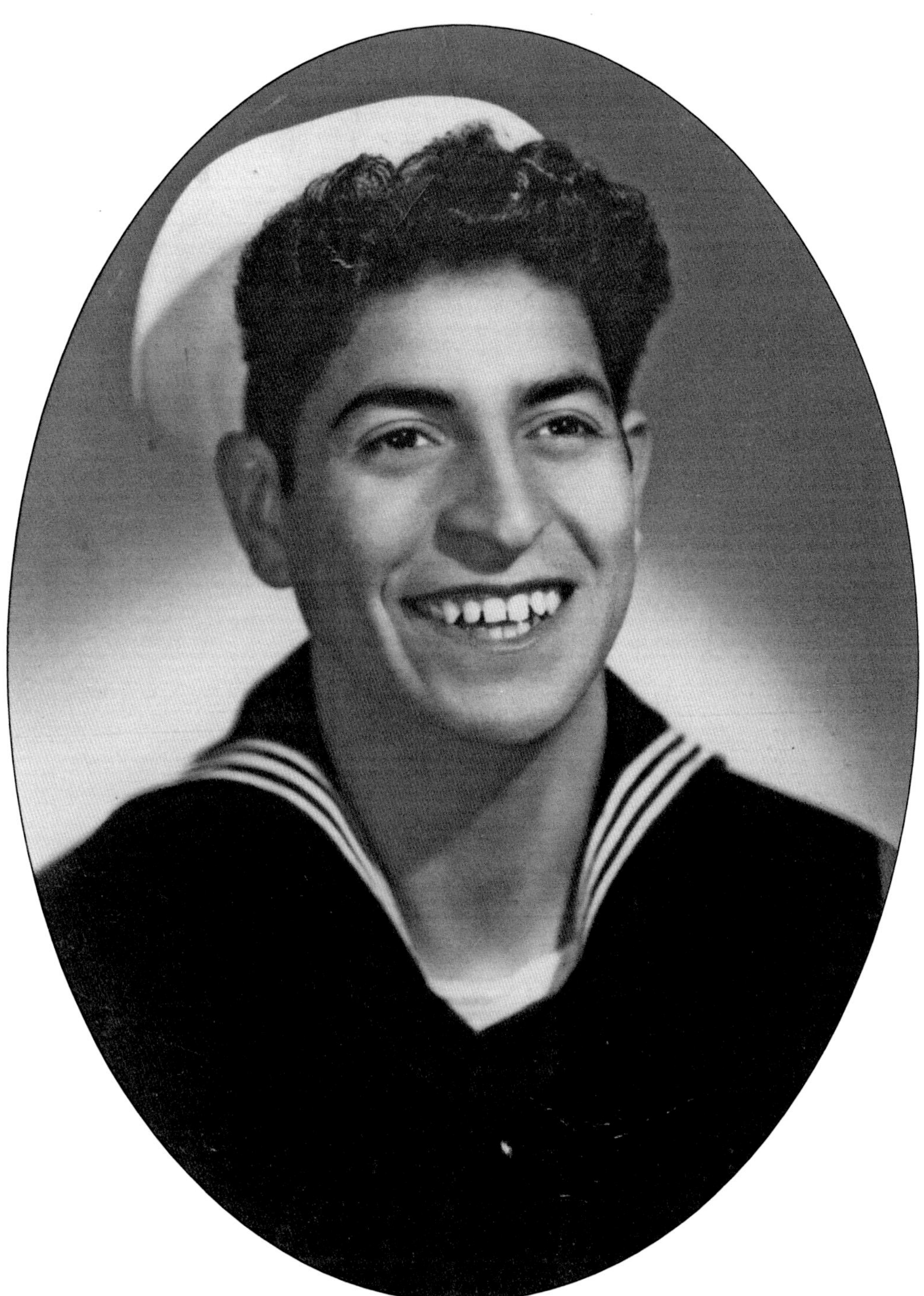

Freddy González is in his Navy uniform. During World War II, there was a shortage of everything and ration stamps were issued according to the size of the family. There were ration stamps for shoes, tires, gas, meat, sugar, tobacco, and other staples. More and more people set up victory gardens with the vegetable seeds the government provided. Others provided their own seeds or plant cuttings to grow vegetables. (Courtesy of González family collection.)

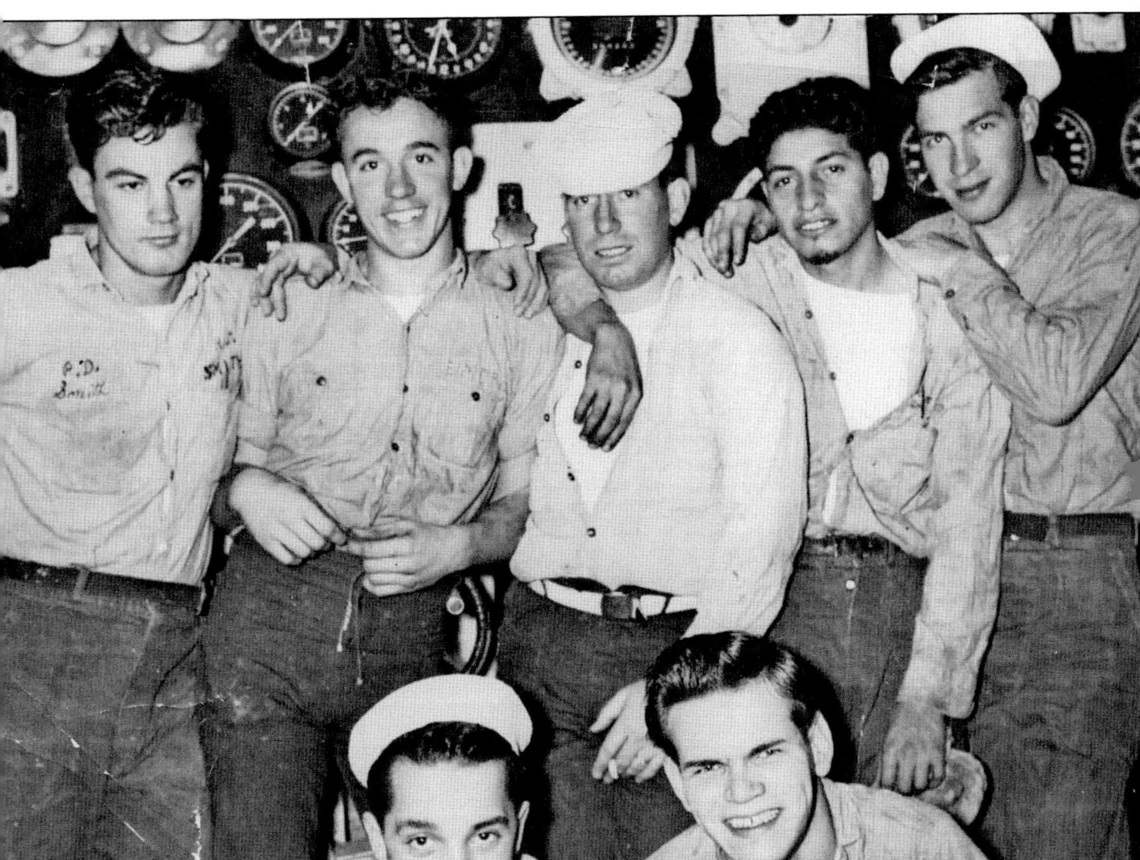

Freddy González is pictured second from right with his military friends. The look of the community changed; like the rest of the country, it adopted camouflage and practiced blackouts in case of night raids. Schools, hospitals, factories, and other large buildings had all their windows painted an olive green to prevent any light from shining through. The roofs were made to look like patches of vegetation instead of buildings. Schools honored the dead soldiers of the community by hanging American flags for those lost in battle. The lists of names posted on the walls kept growing. No football games were played, nor were any yearbooks printed. (Courtesy of González family collection.)

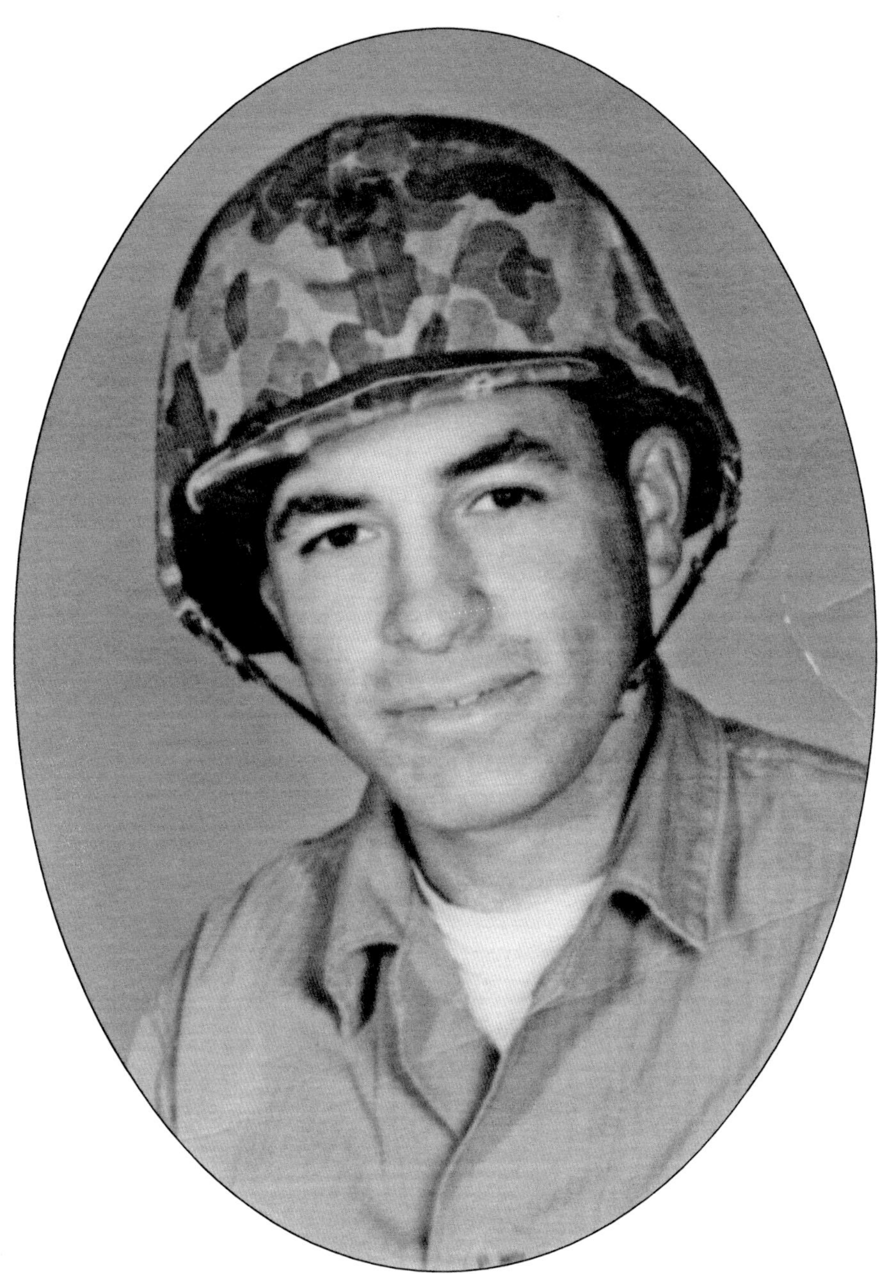

Ernest González is wearing his Marines uniform. Pres. Franklin D. Roosevelt addressed the nation via radio broadcasts, and he would ask the American people to do what they could to keep up morale for soldiers. The entire nation was aware of this, and all did what they could. Rancho Sespe was no different. Ladies baked cookies and other goodies, and teenage girls lined up along the highway to wave to the soldiers in passing convoys. Many fleets of Army trucks with soldiers riding in the back passed along Highway 126 on their way to bases and overseas. The convoys made a sound that could be heard from a distance as they approached. As the trucks rolled past the entrance to Rancho Sespe, many soldiers would throw wadded pieces of paper with their name and address written on them so the girls could write to them. (Courtesy of González family collection.)

Freddy González (left) is seen here with his unidentified Navy friend. The relocation of the Japanese workers also impacted the ranch and its workforce. The loss of manpower brought a migration of families to work on the ranch, and they brought with them new customs and traditions. When the war was over, some went back to doing what they had done before they left, but many boys did not return to the ranch. Some took jobs elsewhere, went to college, or found new lives in other parts of the country. Those on the ranch reflect on some of those changes and talk about how they "grew up" and no longer played around on the ranch as before. The entire country lost its childhood innocence and replaced it with war's gripping realities. (Courtesy of González family collection.)

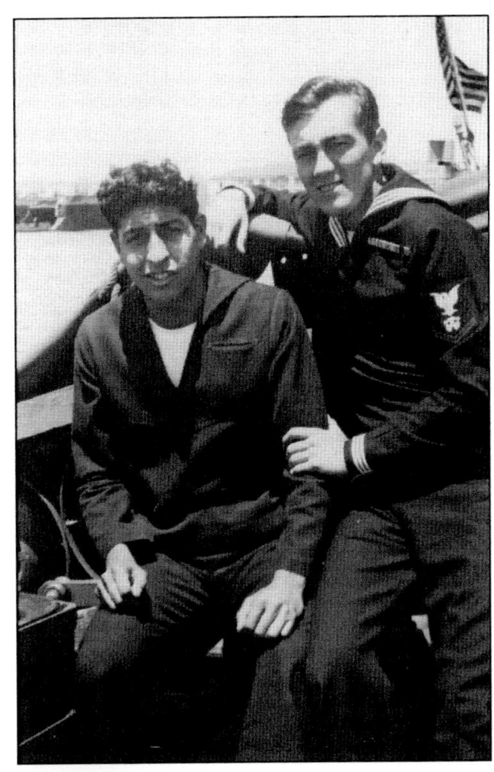

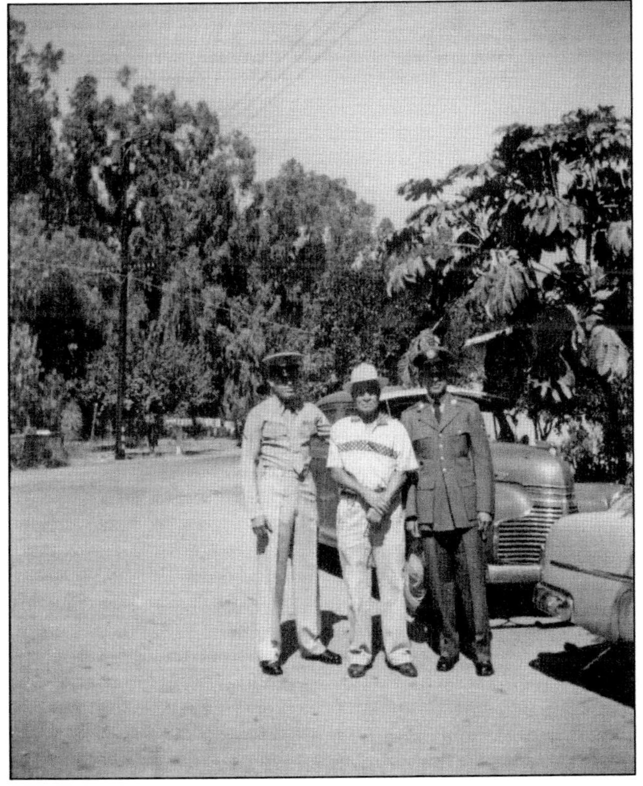

As identified by Becky Morales and the González family, Ernest González is on the left, Federico González in the center, and John "Shorty" González on the right. (Courtesy of González family collection.)

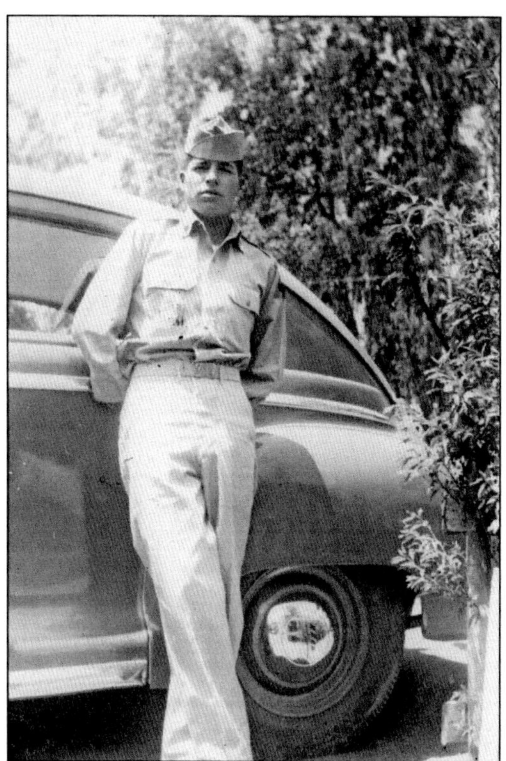

As identified by Becky Morales and the González family, George "Rudy" González was in the Army. His brothers Freddy, Ernest, John, and Salvadore also served in the armed forces. (Courtesy of González family collection.)

As identified by Becky Morales and the González family, George "Rudy" González is in Pusan, South Korea, in this photograph. George told his family and friends that it was very cold in Pusan when he was stationed there. He recalls being cold all the time. (Courtesy of González family collection.)

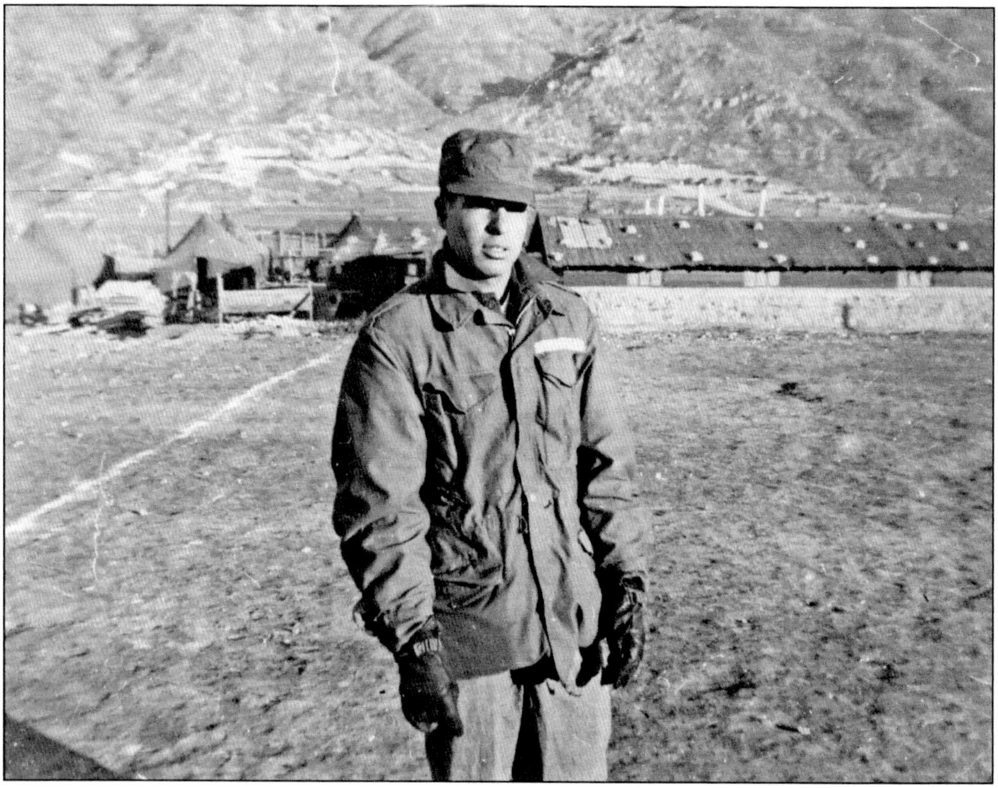

Freddy González (left) is pictured here with a Navy buddy. Freddy enjoyed his friends, and he had a sense of humor, so he made the best of every situation. Freddy was in the Army for three years, then he was in the Navy, stationed in West Virginia. (Courtesy of González family collection.)

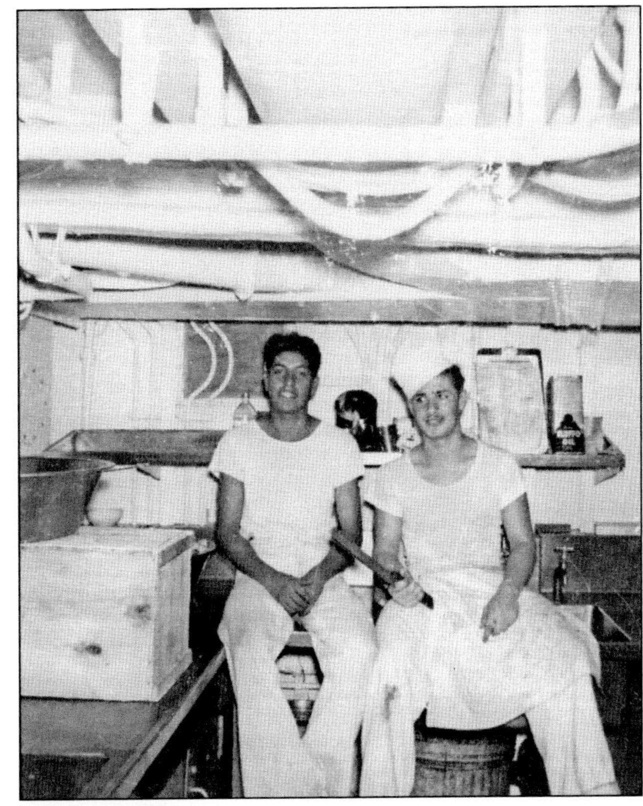

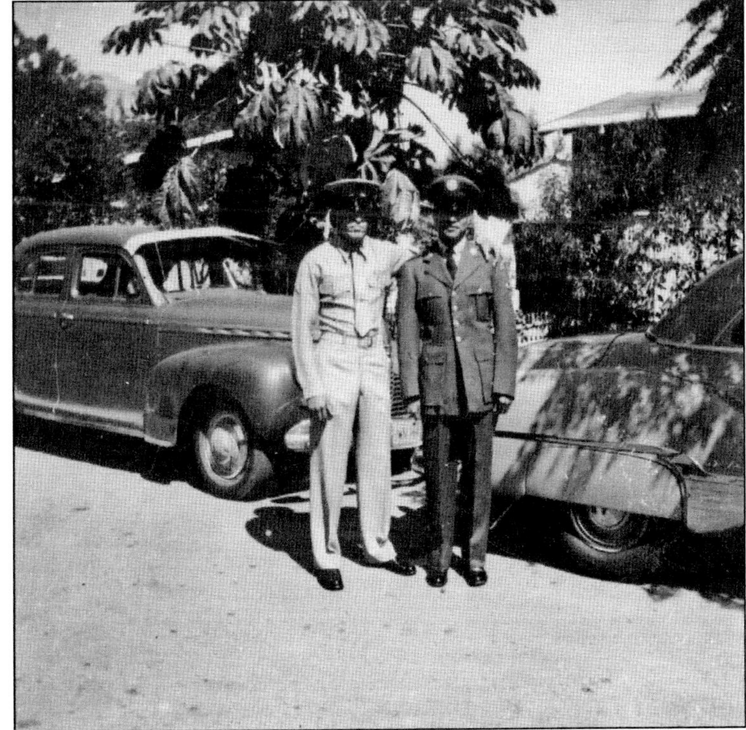

As identified by Becky Morales and the González family, Ernest González (left) and John "Shorty" González pose in their uniforms. (Courtesy of González family collection.)

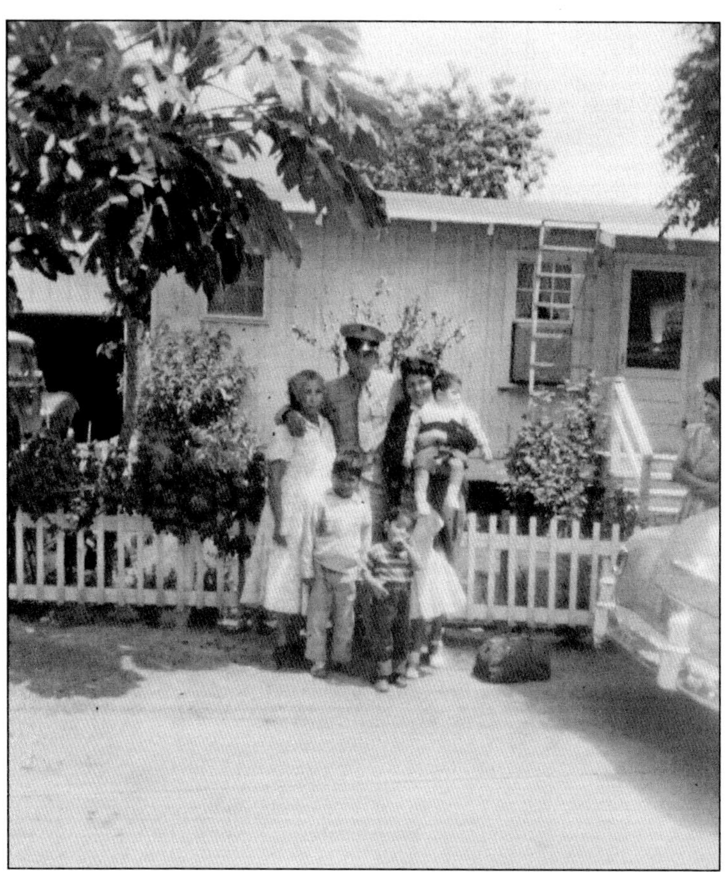

As identified by Becky Morales and the González family, Ernest González poses with his parents, Felicitias and Federico González. While the young men went to war, the boys at home were curious about the German prisoners of war. One day, Bobby Villaneueva and Marvin Kruse went to talk to them. One of the Germans was grumpy and the other POW scolded the boys, but then they smiled at Marvin and Bobby and were friendly. (Courtesy of González family collection.)

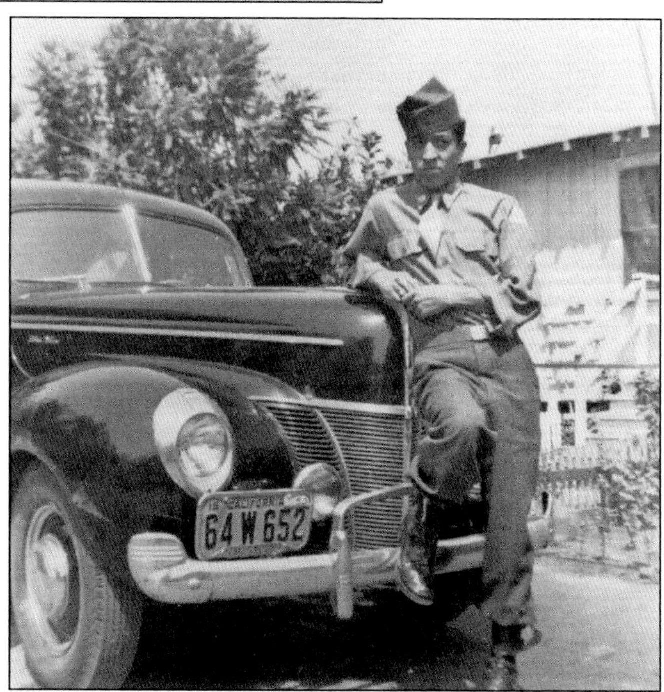

Robert González poses for another picture. Robert was in the National Guard, then he joined the Army. (Courtesy of González family collection.)

George González is shown in 1953 in Korea. (Courtesy of González family collection.)

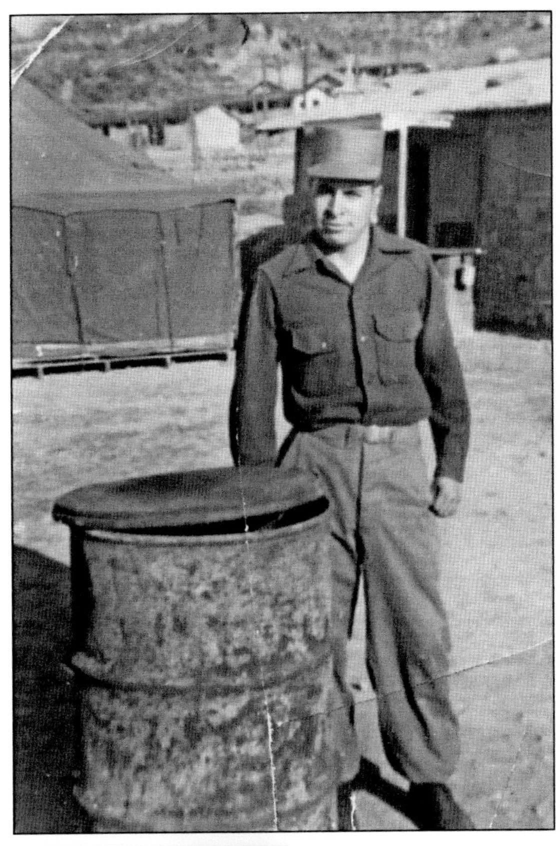

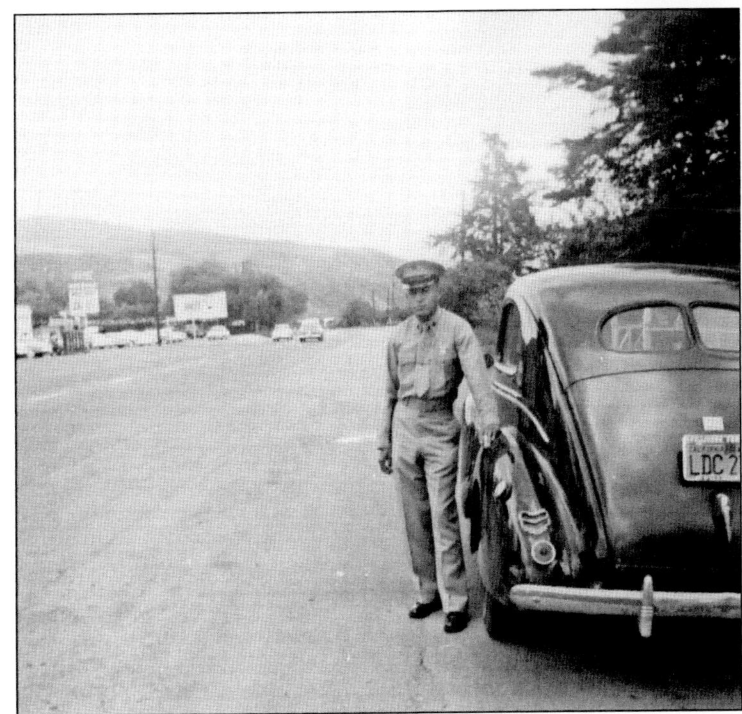

Ernest González is pictured here. He was a marine, and later, he and his wife lived in Fillmore with their eight children. (Courtesy of González family collection.)

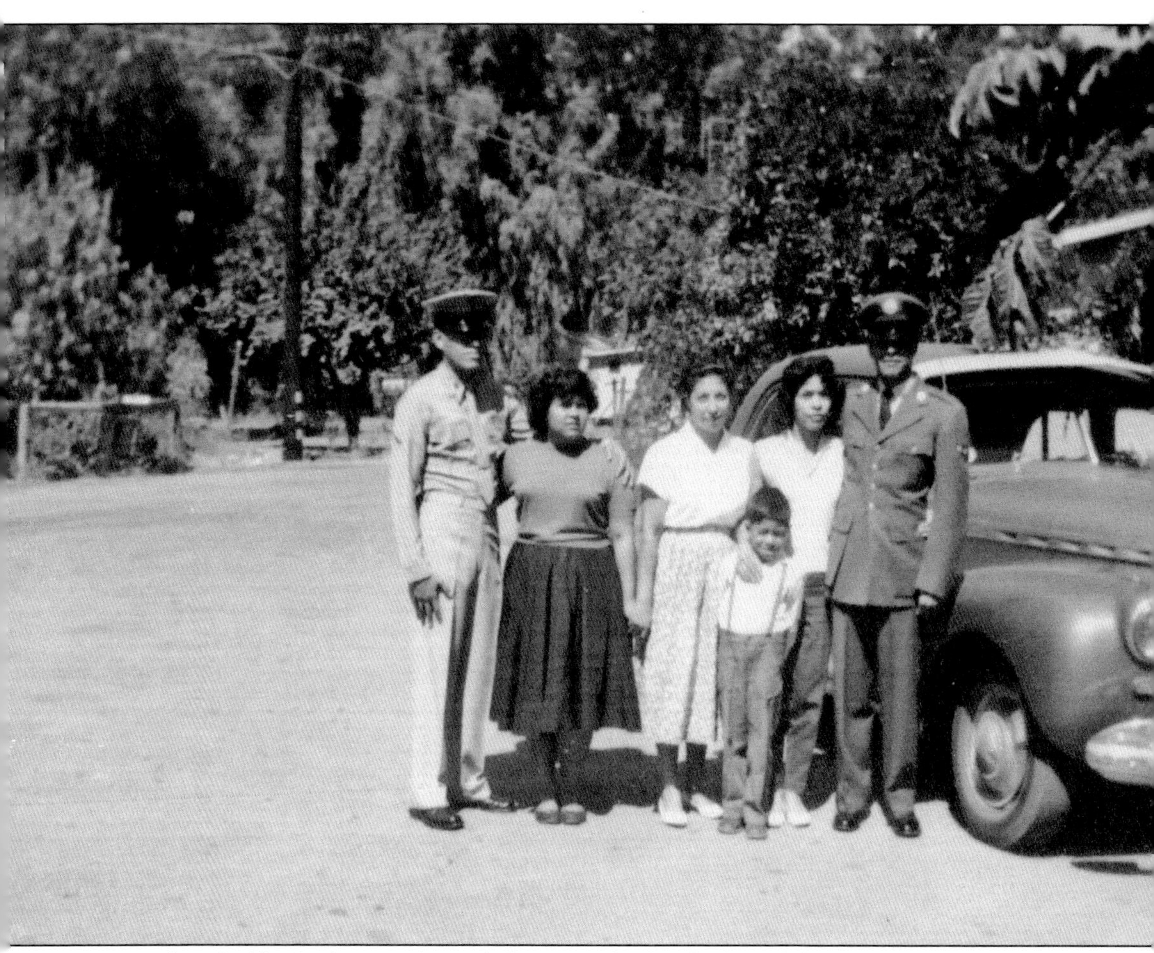

As identified by Becky Morales and the González family, from left to right are Ernest, Connie, Federico, one of Rachel's children, Rachel, and John "Shorty" González. Connie married and had two children. Rachel also married, and she and her husband raised five children in San Fernando. (Courtesy of González family collection.)

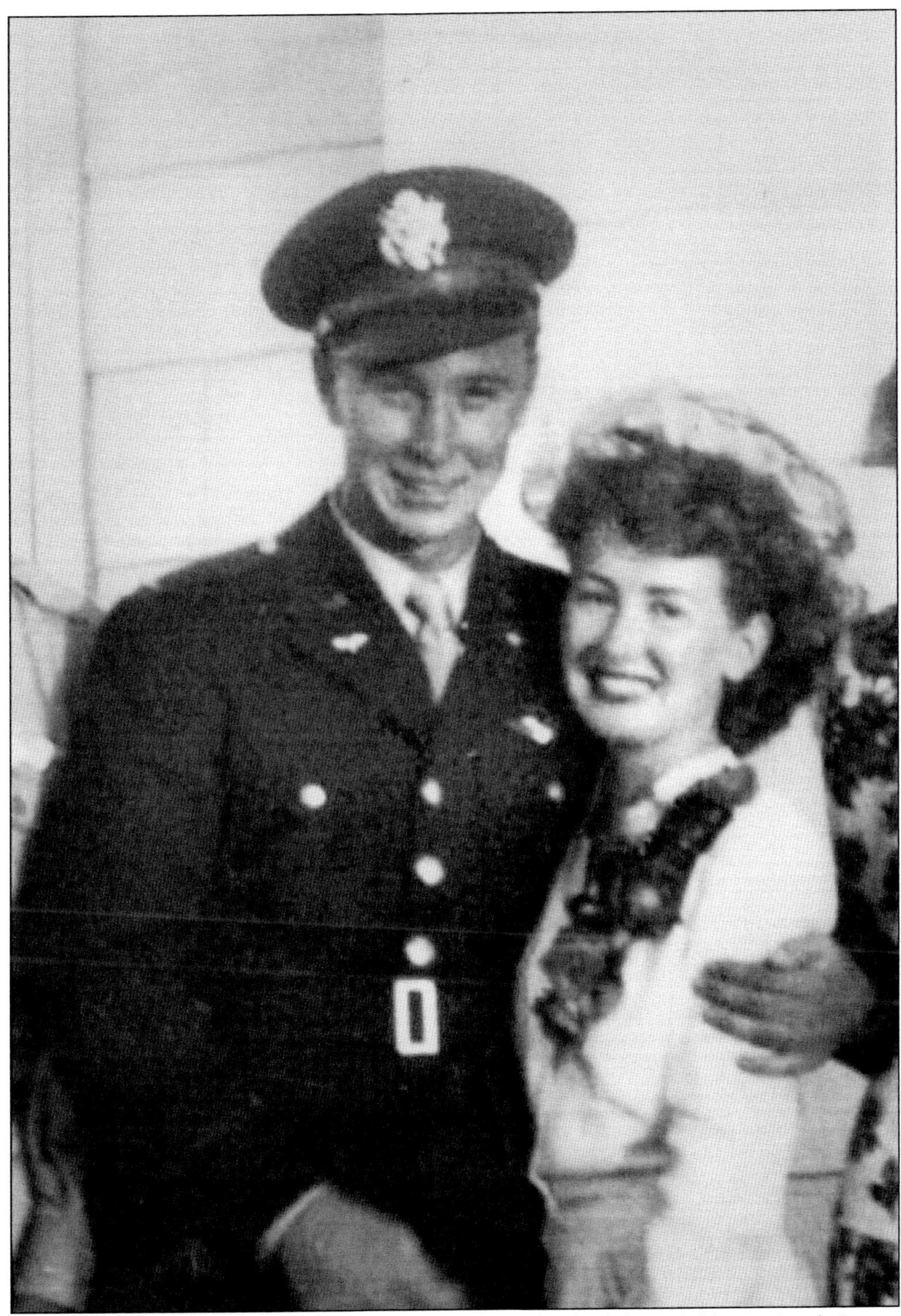

George and Lawanda "Lois" Campbell are shown here. They had three children—Carol, George Richard, and Roger. Roger was elected to the Fillmore City Council and he served as mayor of Fillmore.

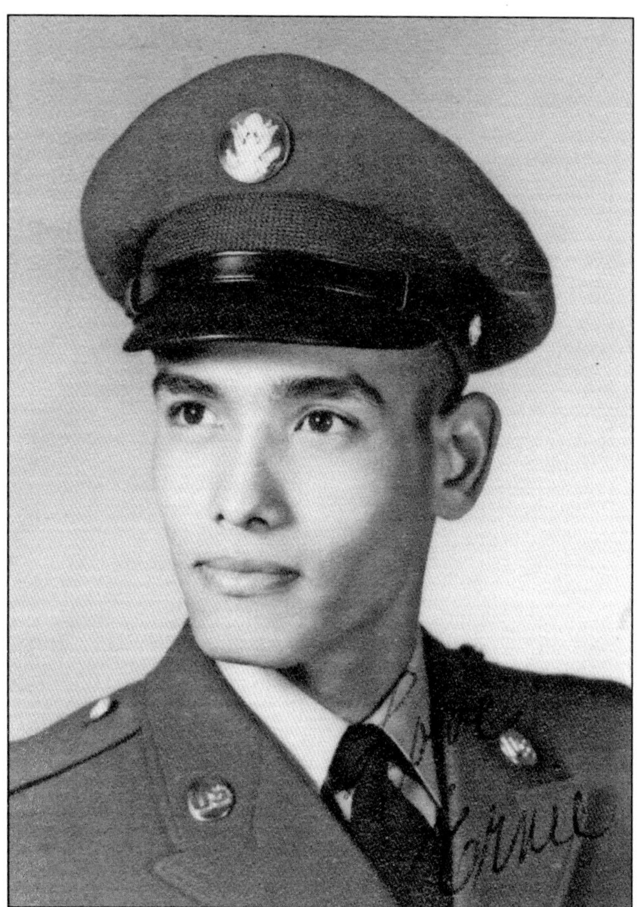

Ernest Morales, pictured in his Army uniform, served on the Fillmore City Council from 1968 through 1984. He was elected mayor in 1973, and again in 1980 and 1984.

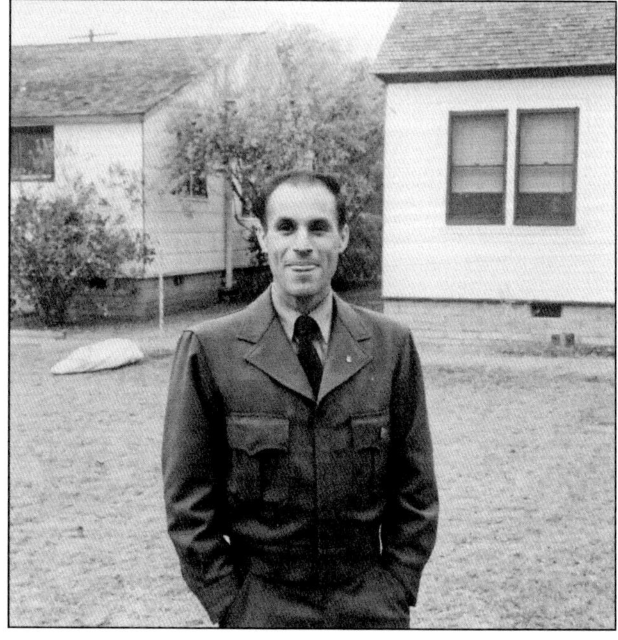

Luther Rodarte is shown at Los Padres National Forest. Luther lost his life battling a blaze in Los Padres National Forest, and he is remembered often by his family and friends. One of the heroes who lost his life serving the public, Rodarte left a widow and five children.

Six

Rancho Sespe and Fillmore in the Modern Age

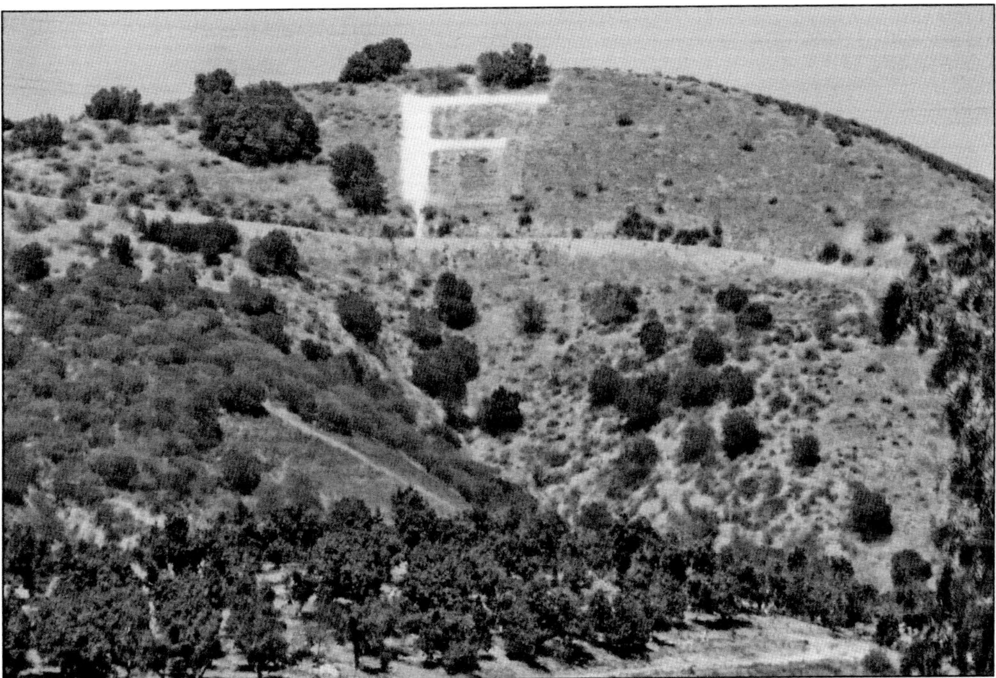

The "F" on the hill was created by three Fillmore boys back in the late 1930s and 1940s. Frank Morales, Leonard Riesgo, and Mike Sanchez carved out the first "F." Since then, it has become a tradition for all freshmen to clean up the area and cover the "F" with lime. T.A. Lombard served as manager of Rancho Sespe from 1947 to 1979 and during this time, orchard acreage increased from 900 acres to 1,800 acres. Twenty-five hundred acres of mountain pasture and river bottomland was used for running 1,000 head of Black Angus cattle. For many years, Rancho Sespe worked in conjunction with the University of California Experiment Station at Riverside to expand the research efforts developed by scientists at the University of California, Riverside. Horticulturists came from all over the world to learn and see the results of these many experiments. (Courtesy of Bob Crum.)

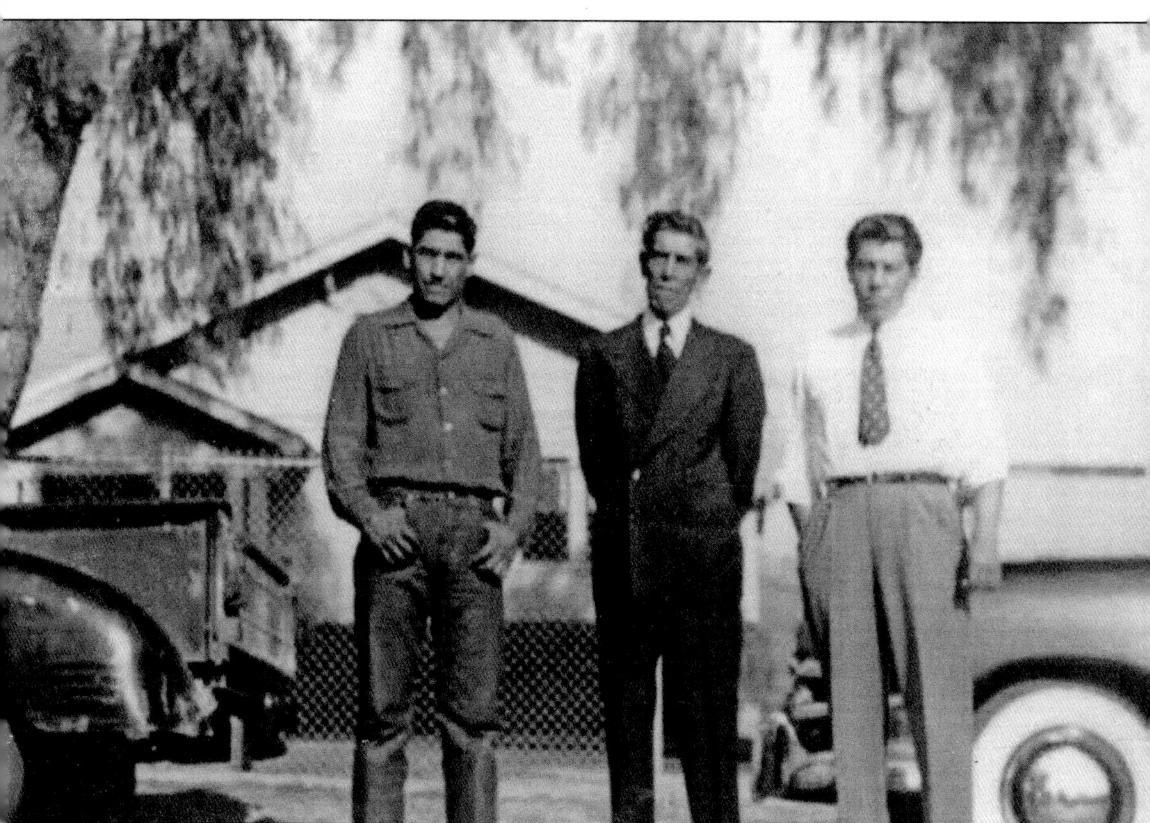

Pictured here, from left to right, are Frank Morales, Cornelius D. Morales Sr., and Cornelius D. Morales Jr. Eudora Spalding passed away in 1942, and she willed Rancho Sespe's 4,300 acres to the California Institute of Technology in Pasadena under a trusteeship. Keith Spalding was to be the trustee until his death. He married his nurse, Lois Fraser, in 1947, and they remained married until his death in 1956. Upon Keith Spalding's death, Herbert Hahn, Esq., became the trustee of the Spalding Trust for another 10 years, at which time the trusteeship ended and the property was deeded to Caltech. The school decided to sell the property to Henry Hillman from Pittsburgh for $6.5 million.

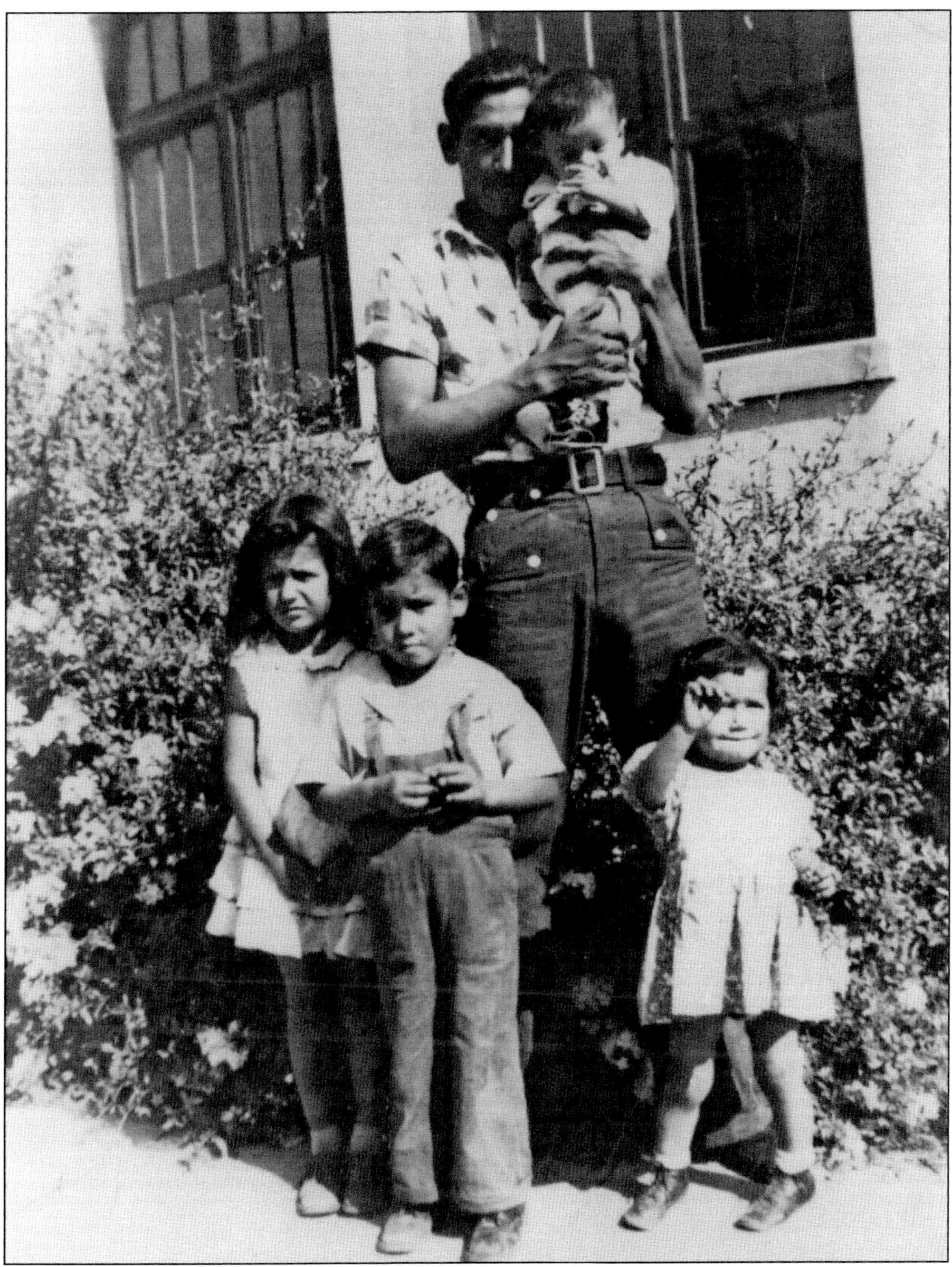

Posing from left to right are Francis, Ernest, Frank, Peter, and Sonia Morales. After several years, Henry Hillman sold the property to Prudential Investment Corporation for $8.5 million. In January 1979, the Prudential Investment Corporation sold Rancho Sespe to a group of San Joaquin Valley growers under the name of Newport Development Company for $11.8 million. The name was changed to Rivcom Ranch.

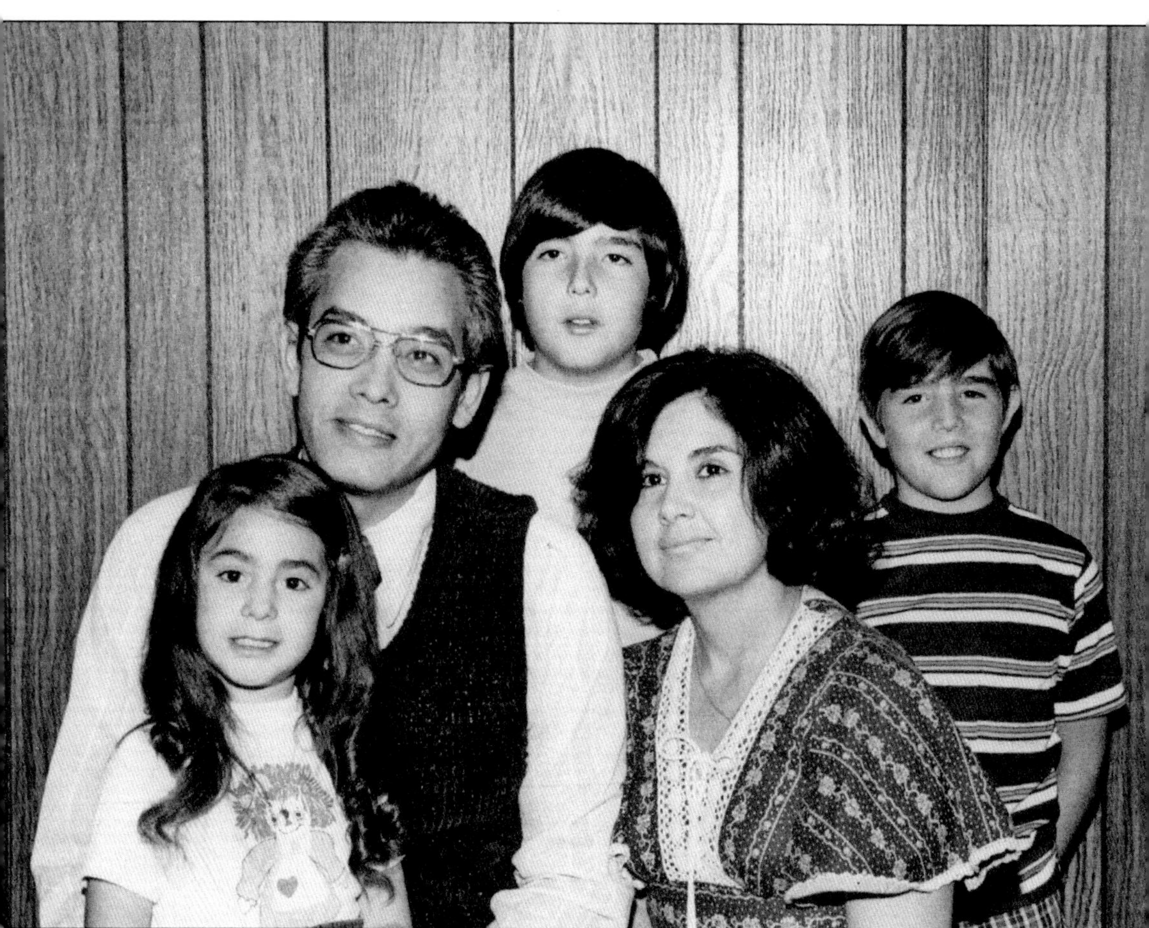

Ernest Morales's family is shown here in 1975. From left to right are Michelle, Ernest, Jeff, Becky, and Brian. The property was next leased to Larry Harris for 10 years. Harris owned a packinghouse in the San Joaquin Valley, so all the fruit was shipped by truck to Sanger. At the end of the 10-year lease, the ranch's 4,300 acres were sold in parcels of 40 acres or more. The zoning for greenbelt agriculture permits only one house for each 40 acres. This brought to a close the single ownership of a large-scale farming operation that was developed over 100 years. When the prime orchard land was split into 40-acre parcels and sold for $20,000 per acre, the total sale price of the 4,300 acres was $27 million.

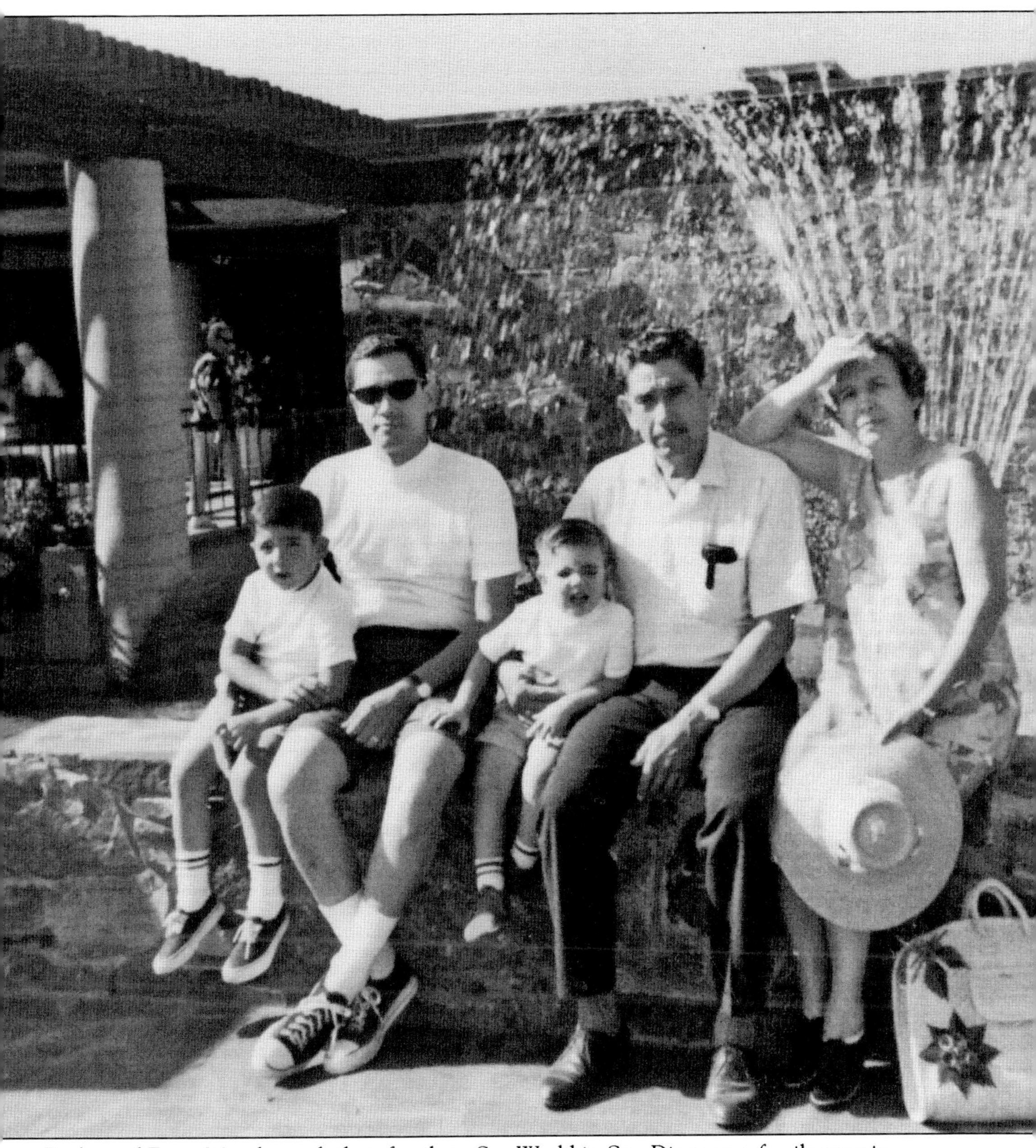

Becky and Ernie Morales took their family to Sea World in San Diego on a family vacation.

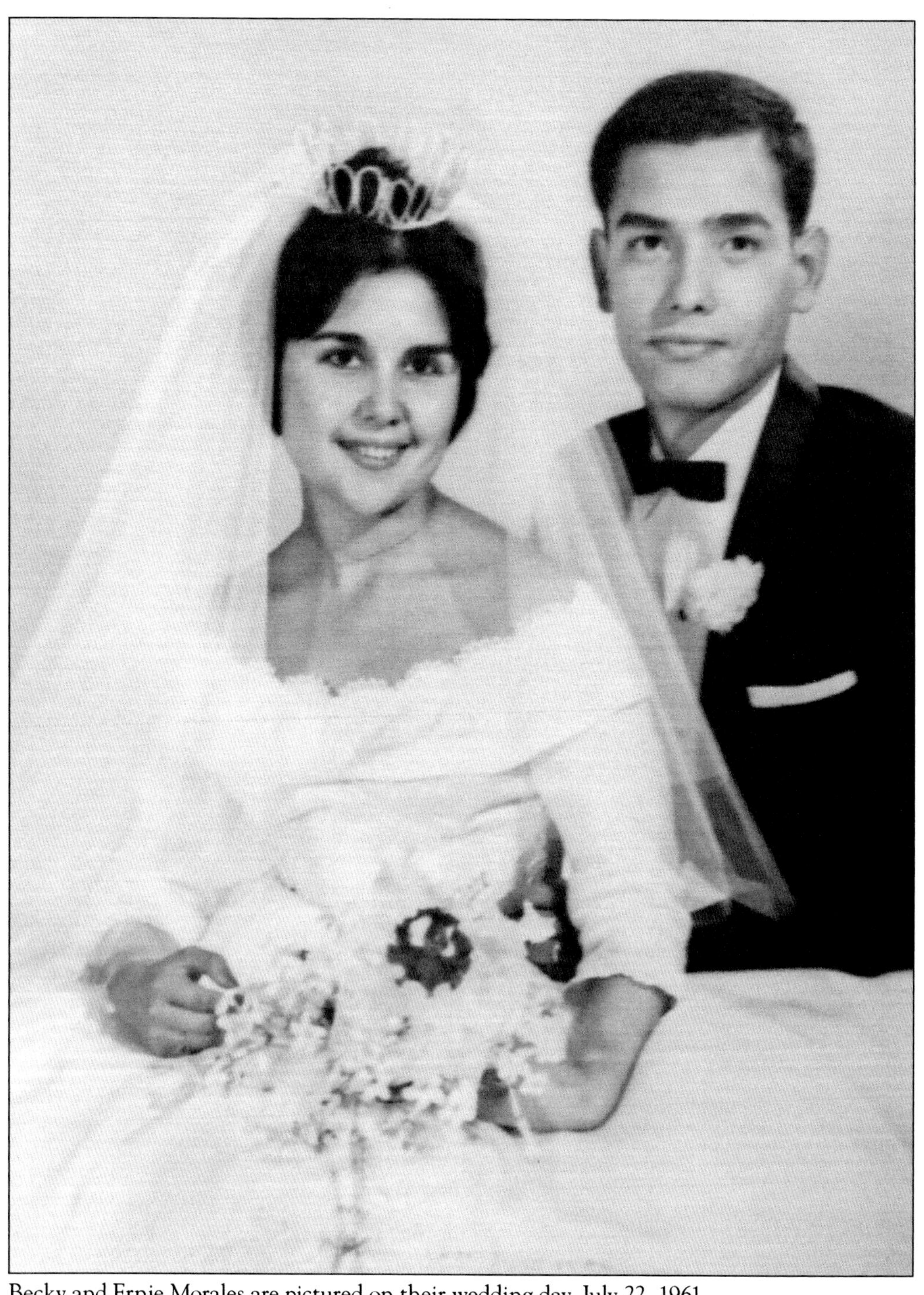

Becky and Ernie Morales are pictured on their wedding day, July 22, 1961.

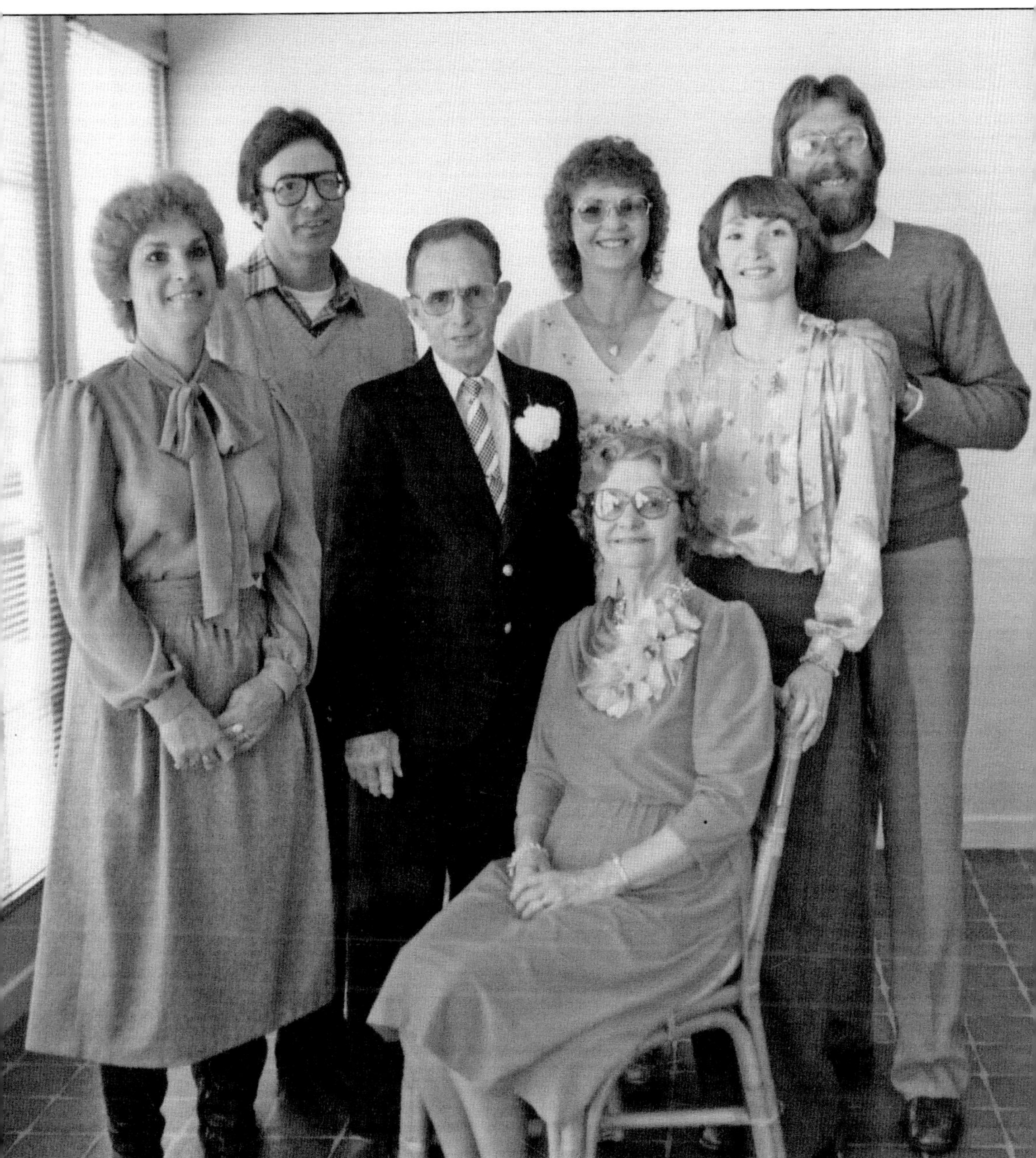

This photograph was taken on the wedding anniversary of Perry and Opal Crockett. As identified by Becky Morales, from left to right are Oma, Dale, Perry, Joan, Opal, Gail, and Jerry. (Courtesy of Gail Crockett family collection.)

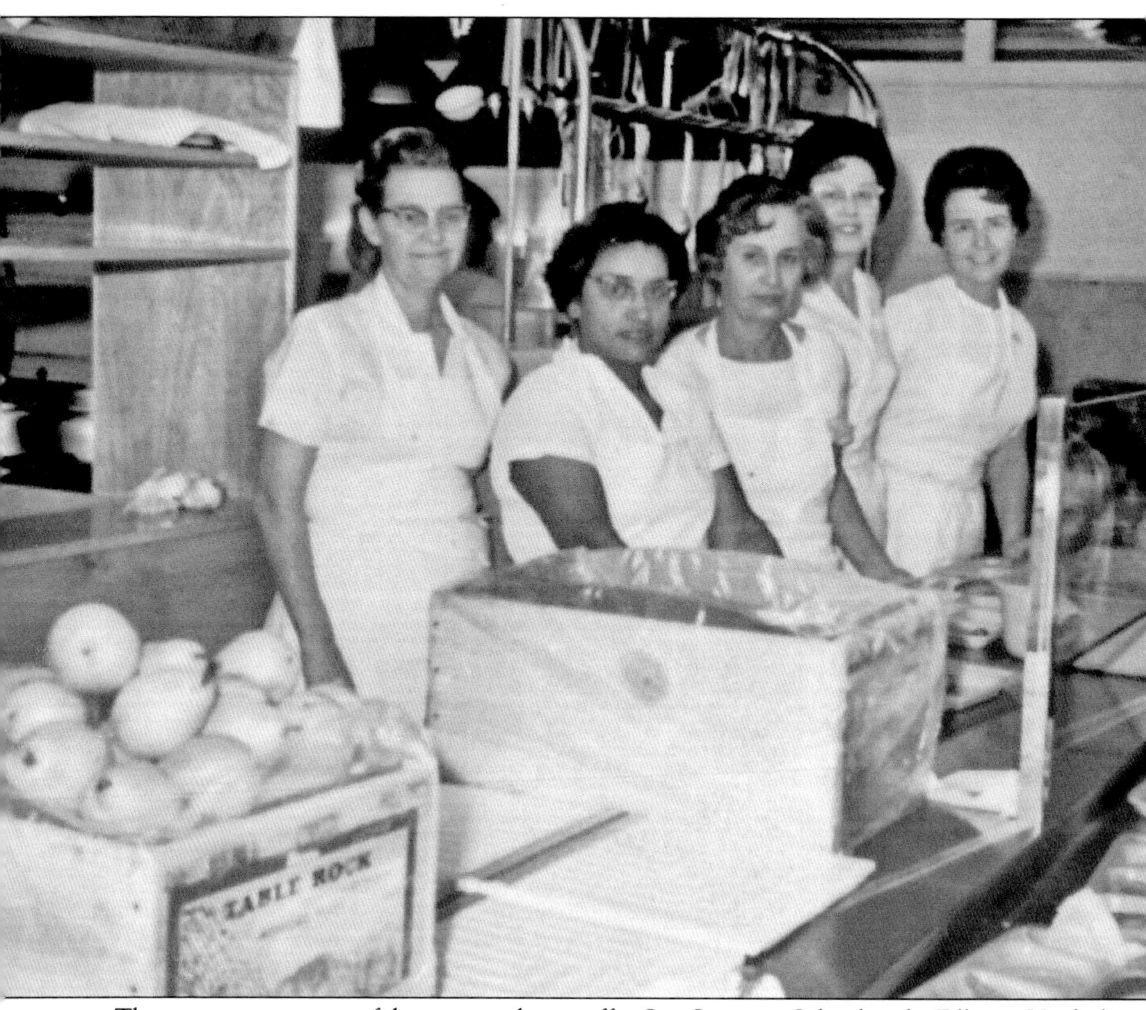

These women were part of the great cooking staff at San Cayetano School in the Fillmore Unified School District. They are, from left to right, Opal Crockett, Carmen Davis, Bea Kaufman, Evelyn Paragien, and unidentified. (Courtesy of Ybarra family collection.)

Hazel Hiberly is pictured on her last day as principal at San Cayetano School. She turns the keys over to Don MacBrine, who was a principal in the Fillmore Unified School District until his retirement. Hiberly used to teach at the San Cayetano School on Seventh Street near Rancho Sespe, then she moved to the Mountain View School. Hiberly became principal at Mountain View then continued as principal at the new San Cayetano School when it opened in 1957. She was also principal at the original San Cayetano School on Seventh Street in Rancho Sespe. (Courtesy of Ybarra family collection.)

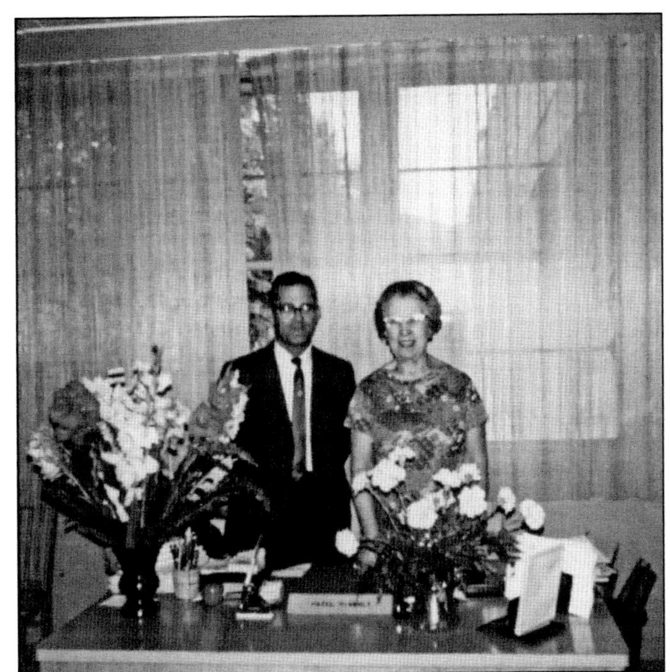

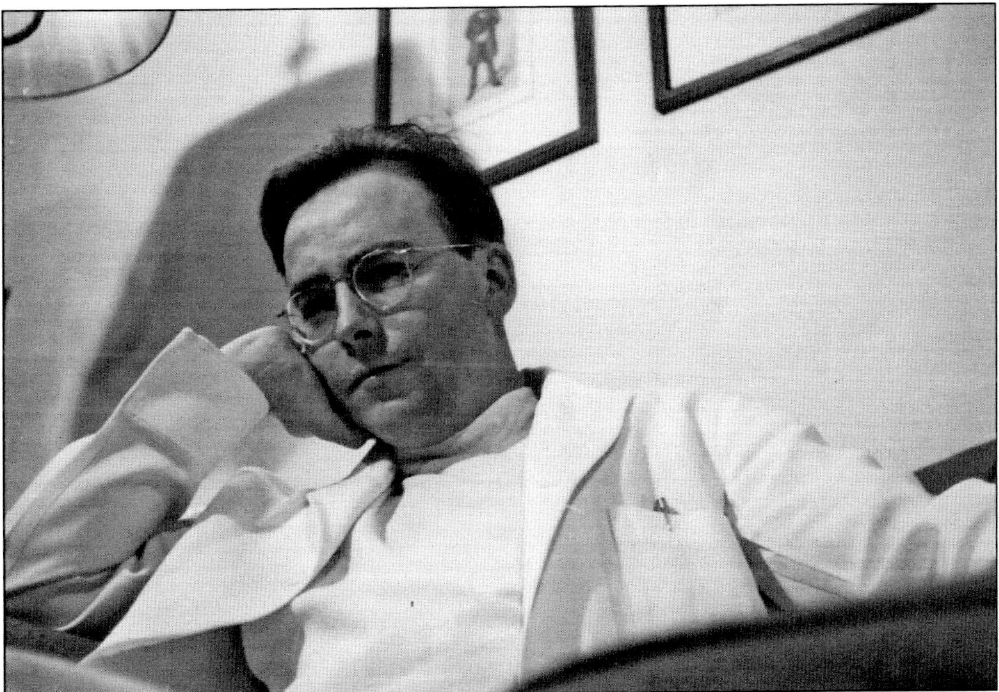

A young Dr. William French is pictured in the early days of his career. He is remembered by many because he made house calls to Sespe Village, where he treated many children when they became ill. Dr. William and Ruth French had three daughters—Penelope, Cathy, and Susan. Another physician in the community of Fillmore, Dr. Donald Musgrave, also cared for many patients. He and his wife, Ruth, also had three children—Ann, James, and Joy. (Courtesy of Susan French family collection.)

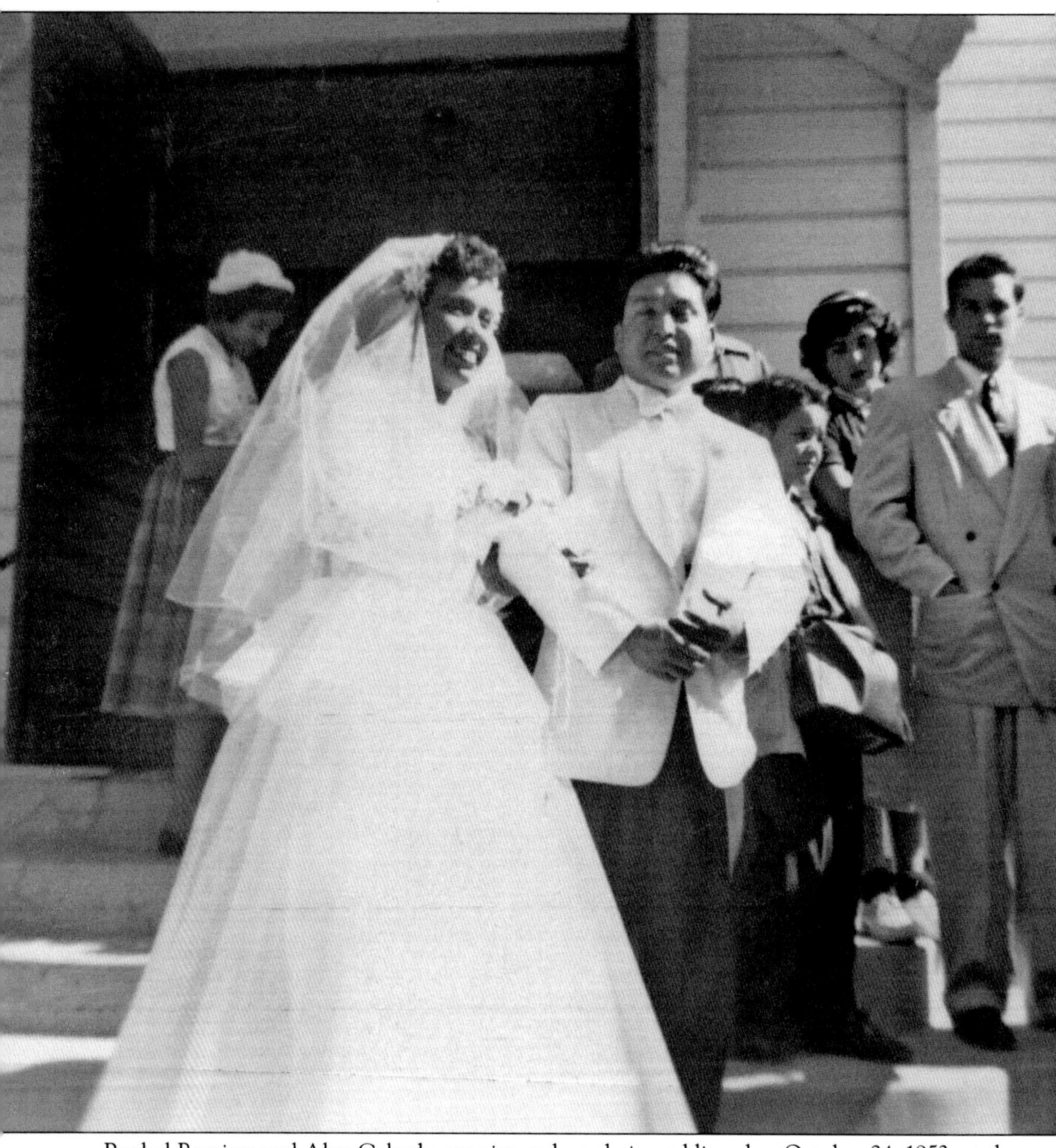

Rachel Ramirez and Alex Calzado are pictured on their wedding day, October 24, 1953, at the St. Francis of Assisi Catholic Church in Fillmore.

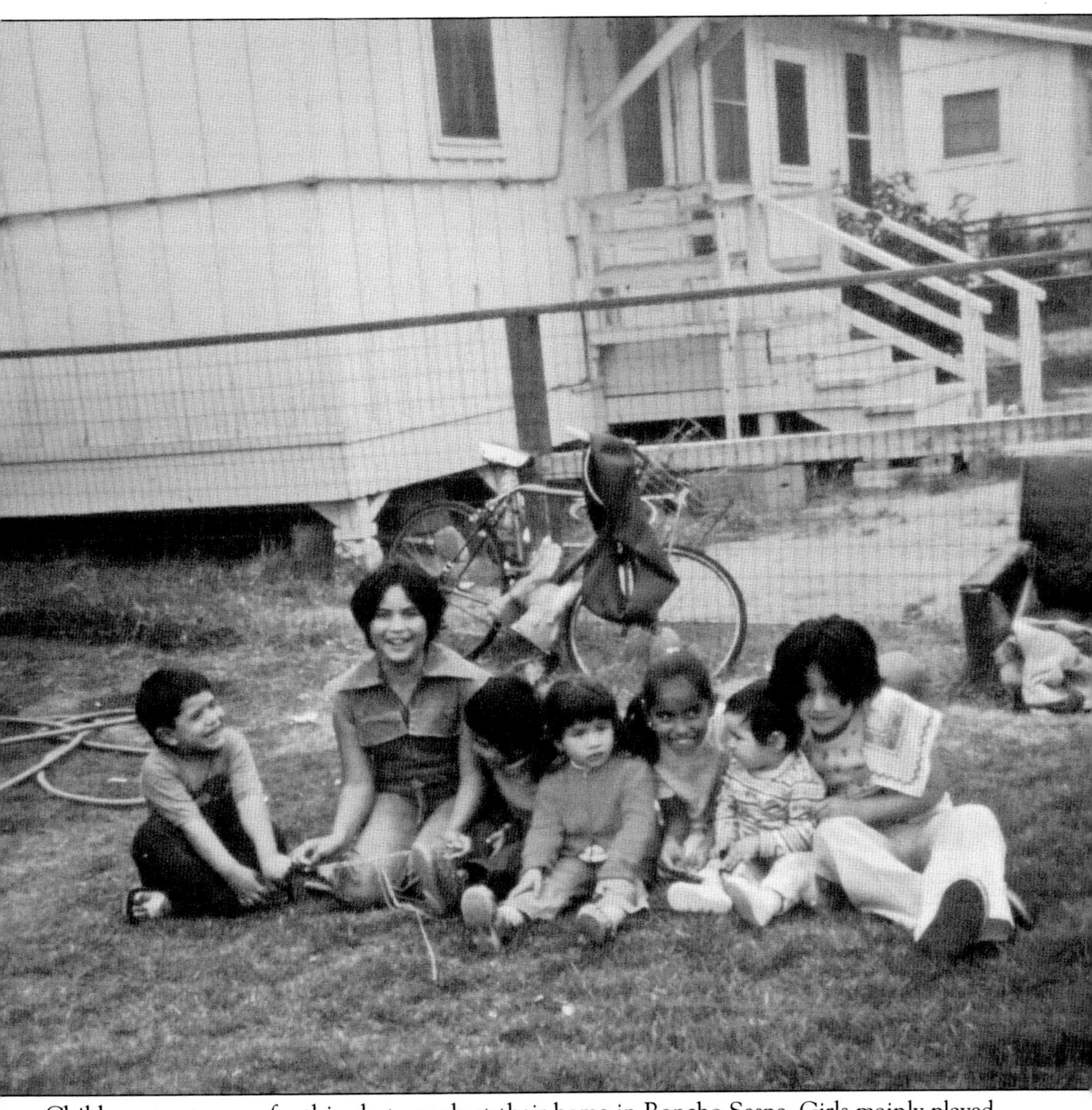

Children stop to pose for this photograph at their home in Rancho Sespe. Girls mainly played jacks, jumped rope, and played with dolls, while mostly boys played basketball or rode go-karts.

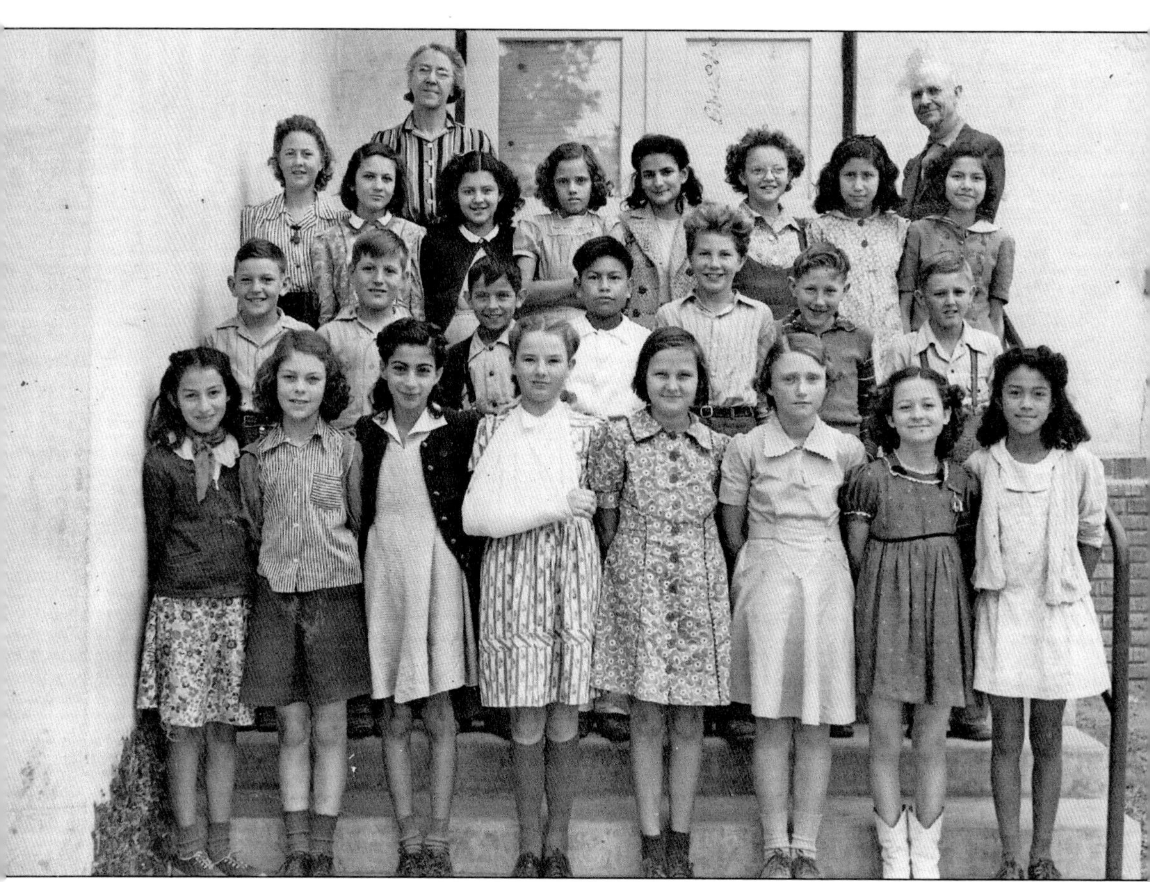

Eva Leon Diaz (front, left) poses with her classmates outside of her classroom at the "new" Sespe School on Sespe Avenue in Fillmore. (Courtesy of FHM.)

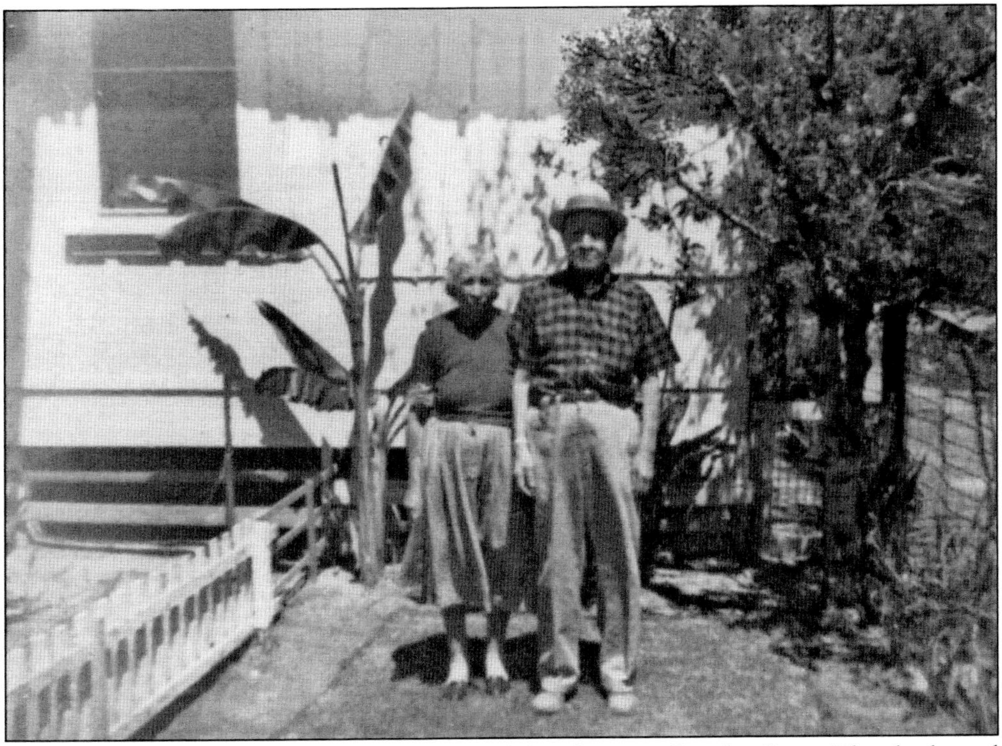

Federico and Felicitas González pose in front of their home in Rancho Sespe. They both used to work for the Spaldings. Federico tended the gardens and did other tasks, and Felicitas assisted Eudora Spalding with the household.

Workers and their supporters are marching with the United Farm Worker flag as a symbol of solidarity. Many wanted to join the union to obtain protection in asking for better wages and work conditions. These were tumultuous times for workers. The public struggles of farmworkers were highlighted through the activism of César Chávez and his followers. They tried to bring attention to the miserable working conditions and the low wages for farm workers. Chávez had roots in Ventura County; he once lived in a home in Oxnard.

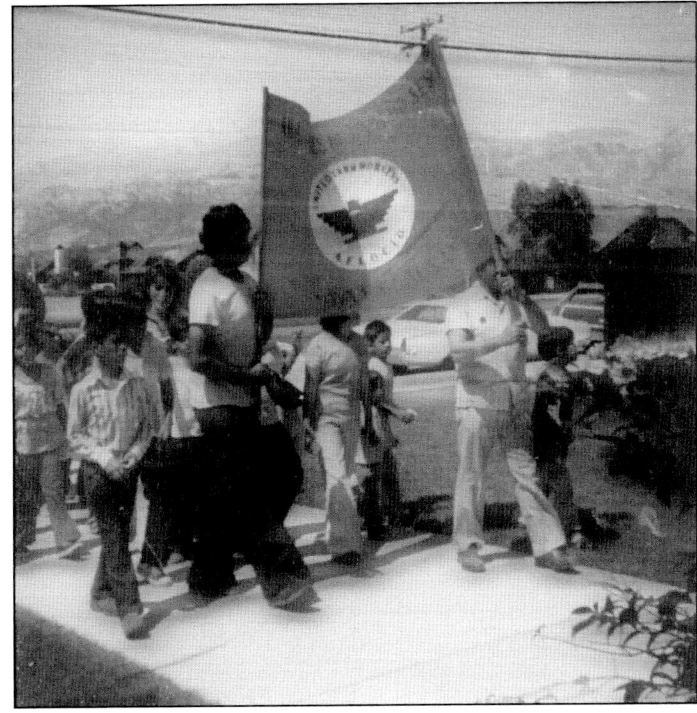

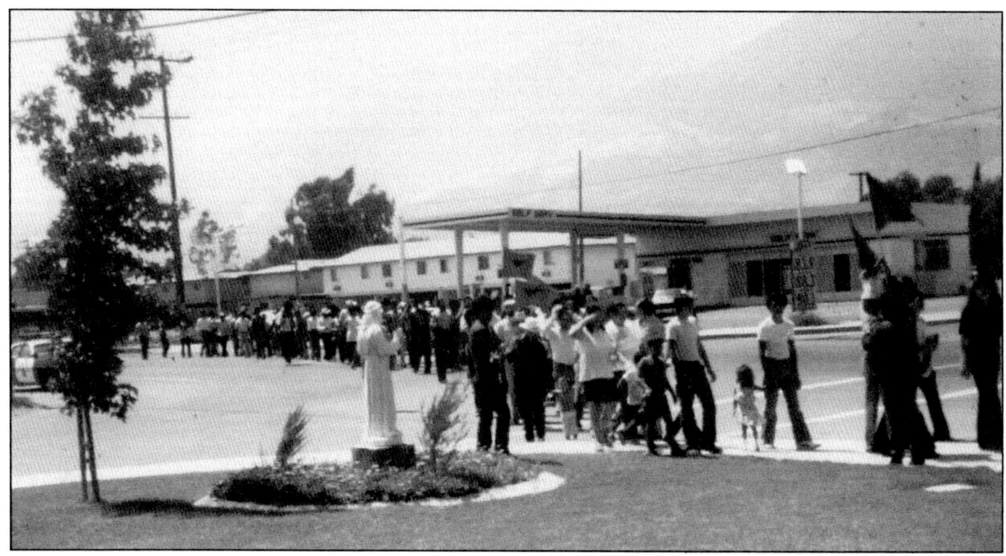

This is another view of the marchers. The farmworkers from Rancho Sespe took to the streets as they attempted to show solidarity with the United Farm Workers, César Chávez's union. Chávez drew attention to his causes via nonviolent demonstrations, marches, boycotts, and hunger strikes. Despite the conflicts Chávez's group had with the Teamsters Union and legal barriers, he secured raises and improved conditions for farmworkers in California, Arizona, Texas, and Florida. Chávez became famous for the slogan "Si, se puede" which is Spanish for "Yes, it can be done."

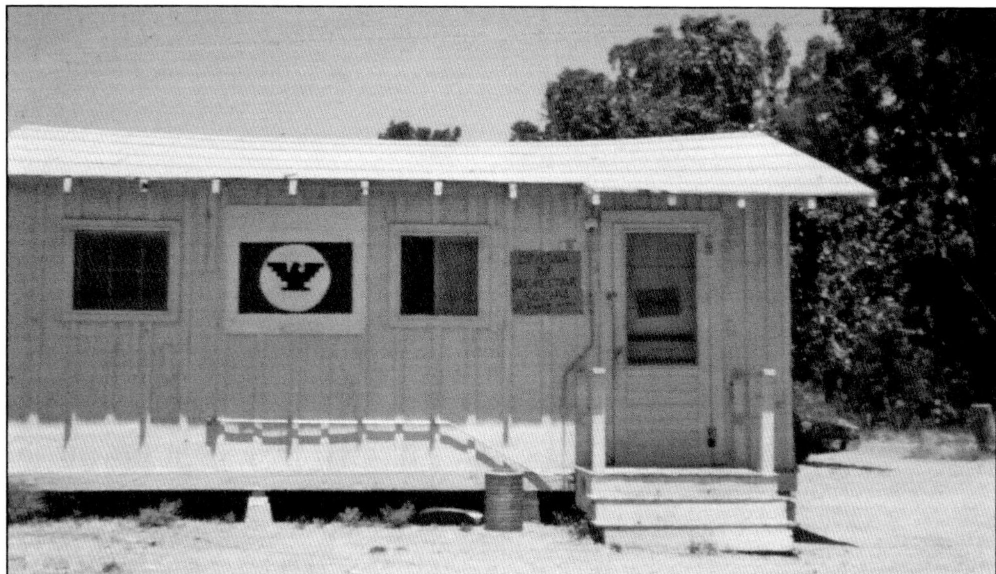

A flag on the side of one of the houses in Rancho Sespe represents support for the United Farm Workers. After 1976, César Chávez led the union through a major reorganization intended to improve efficiency and outreach to the public. In 1984, Chávez inaugurated an international boycott of table grapes because the growers refused to control the use of pesticides on crops. Chávez devoted 30 years of his life to the eradication of the problems of the poorest workers in America. He and Dolores Huerta cofounded the National Farm Workers Association, which became the United Farm Workers Union (UFW) in 1962. March 31 is observed as César Chávez Day in the United States.

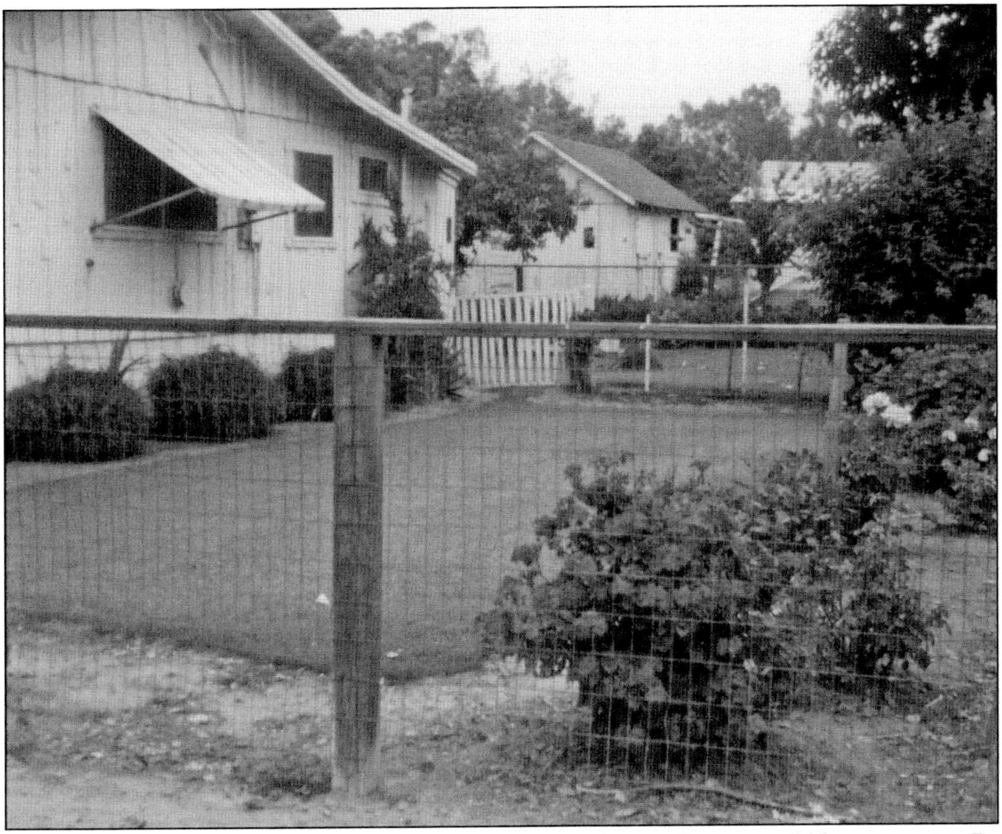

A home in Rancho Sespe has flowers and green grass, but its future would be uncertain. By Kenneth Glenn's accounting, in addition to the two labor camps on the ranch, there were 25 homes scattered throughout the property for supervisory employees. The ranch required a full-time plumber, two carpenters, and two painters to maintain the housing. The workers were charged $15 per month for rent and $4 per month for gas. The water was free. Each house had an electric meter, so each family paid the electric bill. The supervisorial employees paid $25 per month rent and their utilities. In the early years of development, the water for irrigation was pumped by a 100-horsepower steam-powered pumping plant near the Santa Clara River. The water was pumped to a reservoir on the hill by the Spalding house, and it was distributed by gravity to the various orchards by underground pipelines. As acreage of new orchards increased, additional wells were drilled in six locations, and two more reservoirs were built on higher ground to take care of the foothill acreage. All the wells and reservoirs were connected by miles of high-pressure lines.

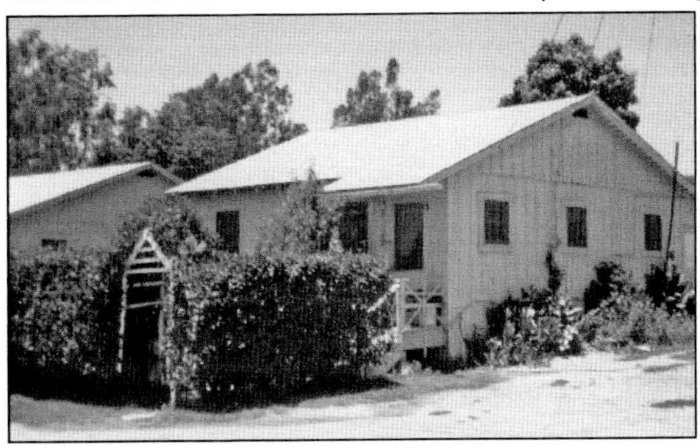

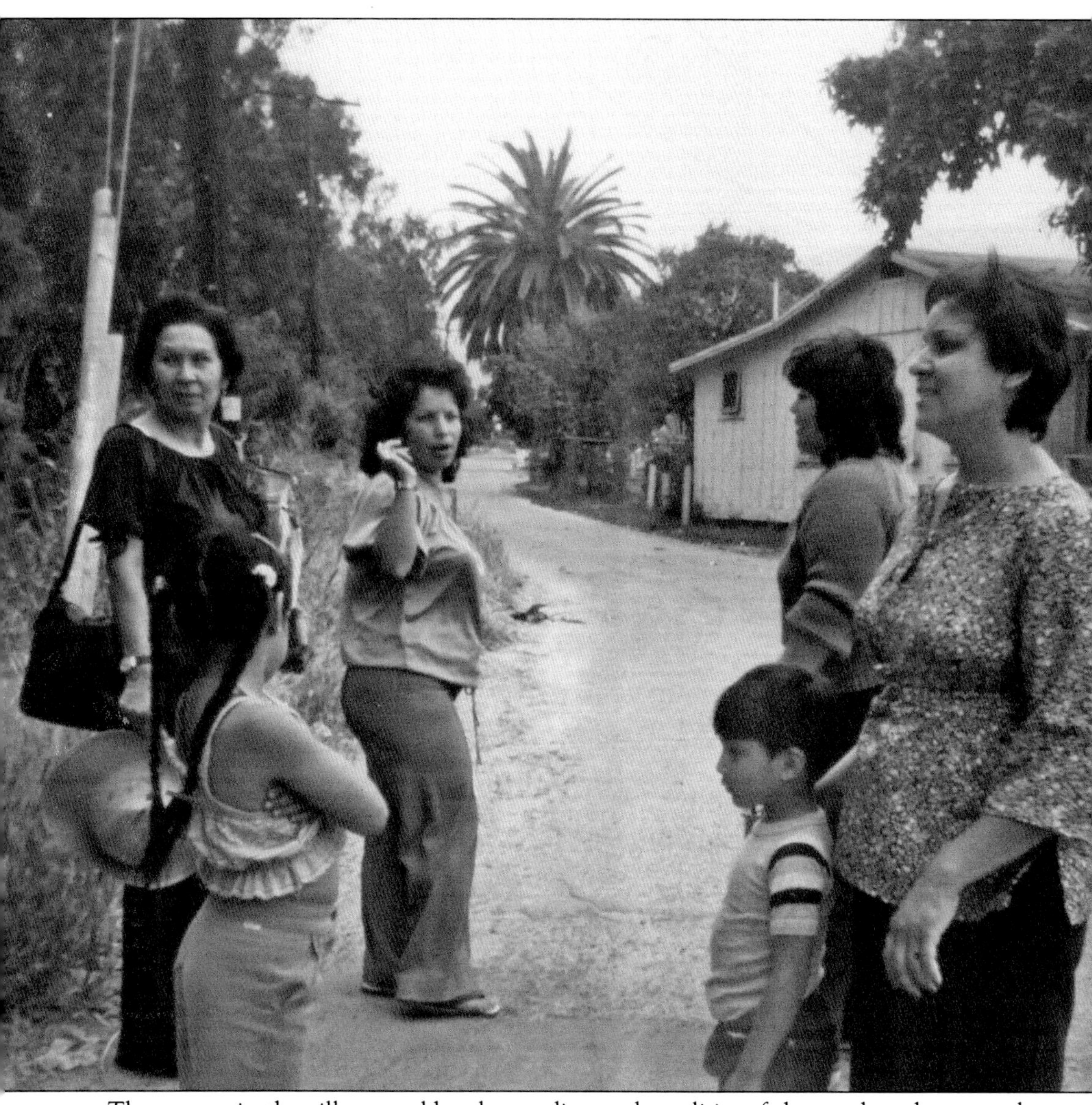

The women in the village would gather to discuss the politics of the ranch and contemplate an uncertain future—what would they do if they had to move away, where would they move, and what jobs would be available to their husbands and family? It was a difficult time, and the changes for Rancho Sespe were not welcomed by those who lived and worked on the ranch. The community of Fillmore struggled with a resolution that the Fillmore City Council passed in April 1985 known as the English-Only Resolution, which stated that the "City of Fillmore recognizes English as its official language." Considered racist, this divided the community. The resolution was repealed almost 15 years later, on December 16, 1999.

The ranch was sold in 40-acre parcels, and the houses were demolished. Here is an example of an empty building during this time.

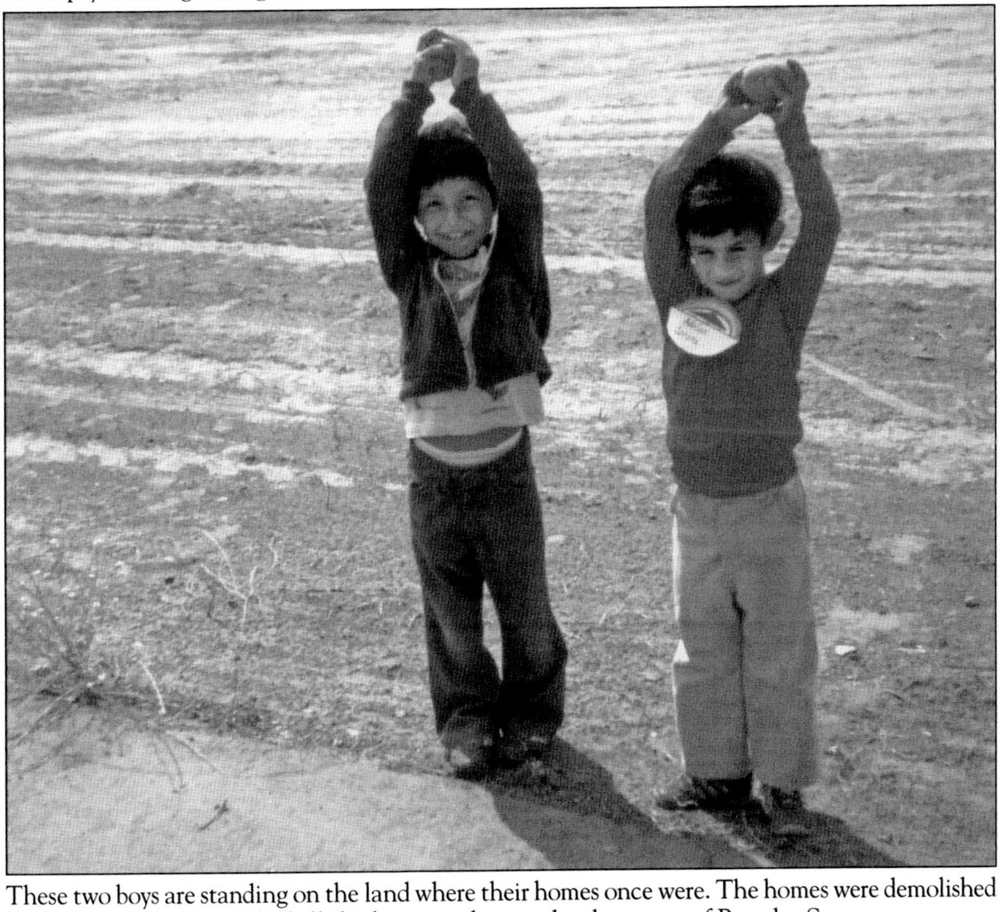

These two boys are standing on the land where their homes once were. The homes were demolished and the land was cleared of all the housing that used to be a part of Rancho Sespe.

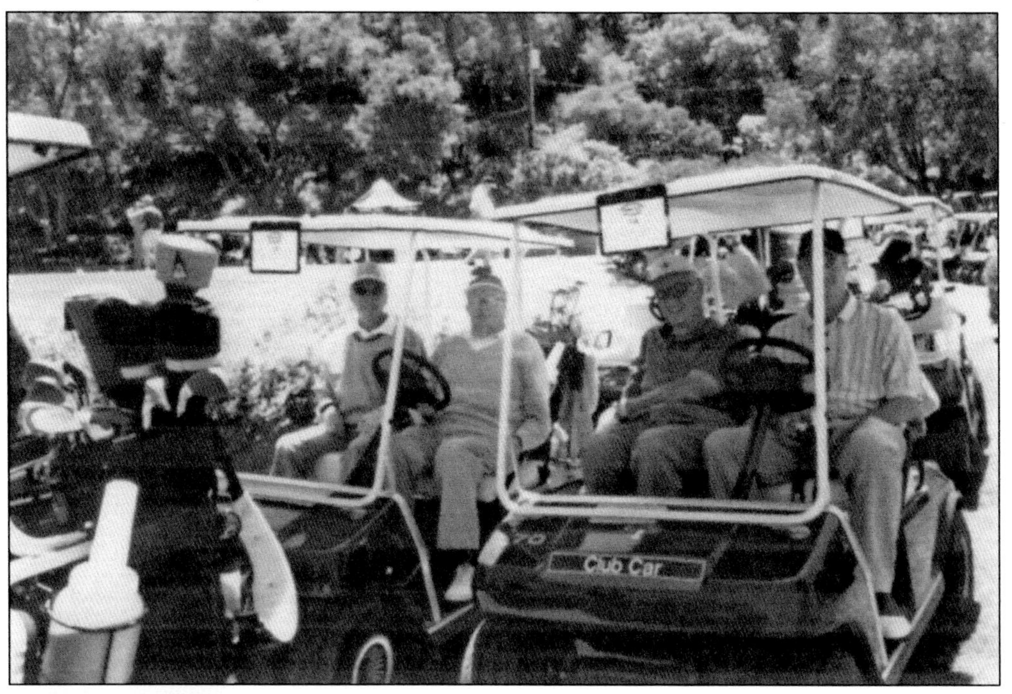

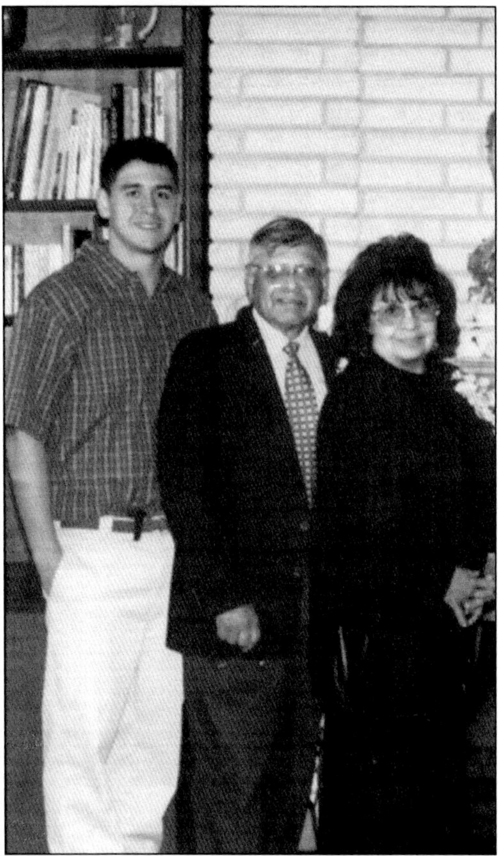

Glen Phillips and Hank Data were the best of friends. They used to play golf all the time and went to Chavez Ravine to see their favorite baseball teams play the Los Angeles Dodgers. Phillips was a generous and kind man. According to the March 4 and March 11, 1938, editions of the *Fillmore Herald*, when Phillips was principal at Mountain View School in Fillmore, a heavy storm washed out the Sespe bridge. Ninety children from Rancho Sespe were stranded in Fillmore, and Phillips took charge and arranged for all the children to be housed with volunteer families until they could safely return home. That evening, he took the students to the movies as a special treat. (Courtesy of Floreine Data family collection.)

Marcelina "Woody" Ybarra attended a game between the New York Yankees and the Los Angeles Dodgers with Glen Phillips and Hank Data. They had front row seats right behind home plate, and they each shook Mickey Mantle's hand. It was a delight to see Woody's face light up whenever he told that story. He is pictured here with his grandson Robert Grosfield Jr. and Minne Ybarra. (Courtesy of Ybarra family collection.

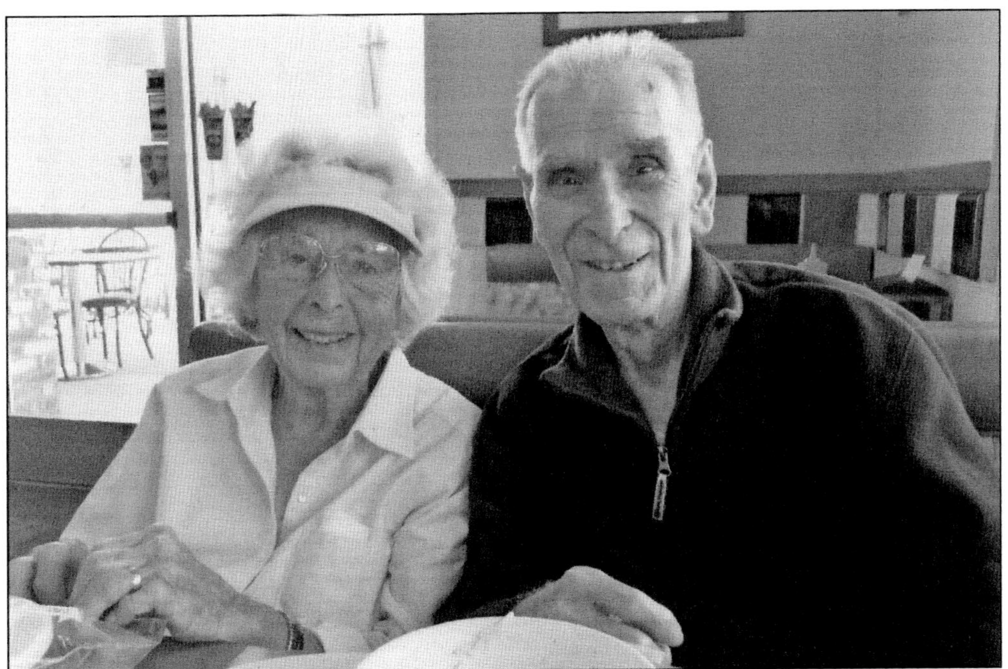

Floreine and Hank Data spent a lifetime together. Hank used to work for Jones Brothers as the parts manager, and Floreine worked in the payroll department for the local growers.

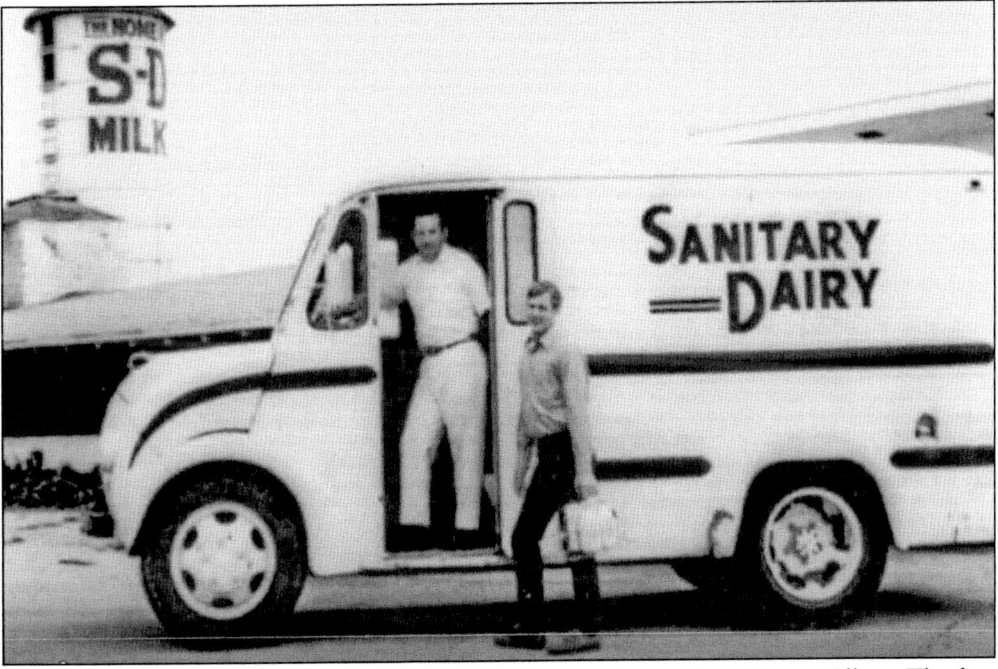

The Sanitary Dairy delivered milk and other products to the people of Sespe Village. The first dairy to deliver there during the 1930s was the Orange Top Dairy from El Rio. Pictured are Paul Conaway and high school student Wayne Sampson, who assisted Conaway. Several high schoolers worked at the Sanitary Dairy, including Rob Scherzinger, a founder of the Aspen Helicopter Company who continues to fly today. (Courtesy of the Russell Hardison family collection.)

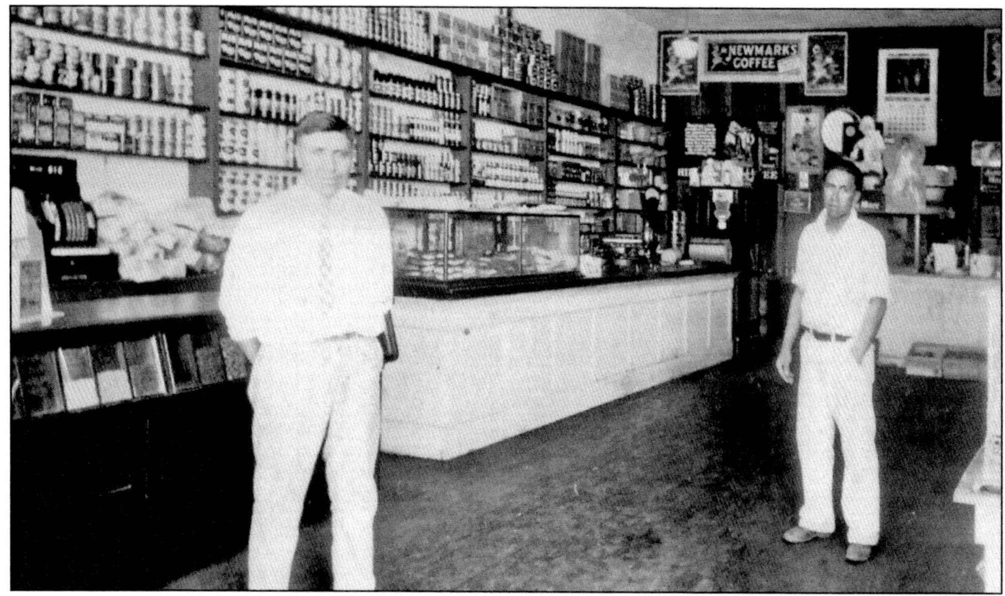

One of the markets in Fillmore used to go out to Rancho Sespe to sell meat and other necessities to the families there. Here is a photograph of the Gonzálezes' La Mexicana Market, which was owned by Jessie González after her husband, Jose, passed away. Her son Richard Sr. helped her and took over the store. His brother "Popeye" and son Richard Jr. assisted with the market. Other markets that might have done business with the people in Sespe Village were the Inadomi Market and the Sanchez Mexican Mercantile Market. The Inadomi Market was owned by John Inadomi; Abundio Sanchez owned the Sanchez Mexican Mercantile Market.

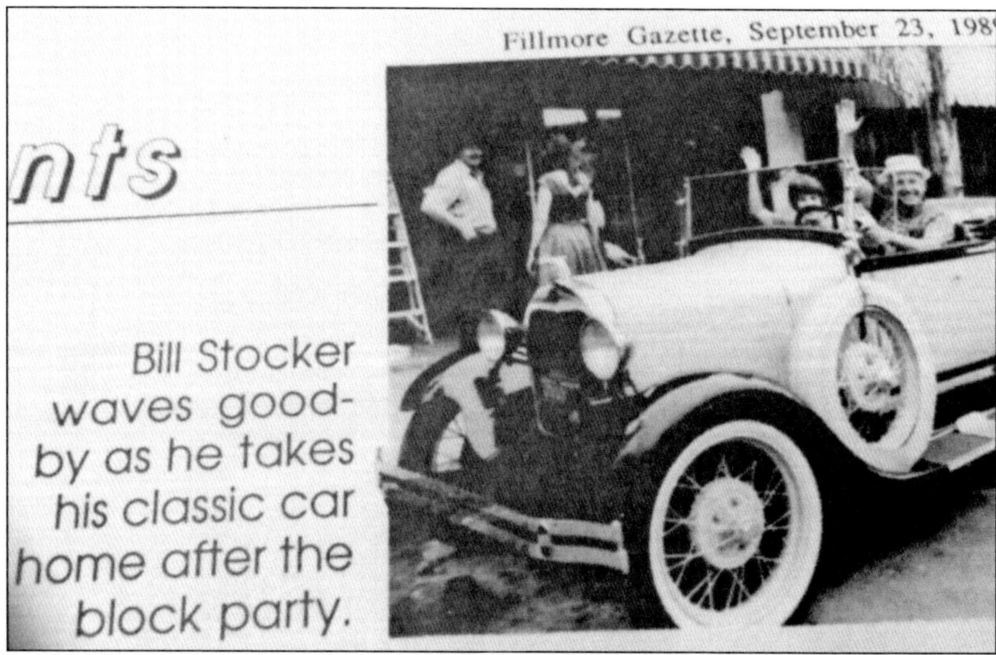
Fillmore Gazette, September 23, 198

nts

Bill Stocker waves good-by as he takes his classic car home after the block party.

Bill Stocker owned Stocker's Department Store for many years. Here, he is enjoying one of Fillmore's traditions, taking part in the Fillmore Festival Parade. Bill and his wife, Pat, still reside in Fillmore.

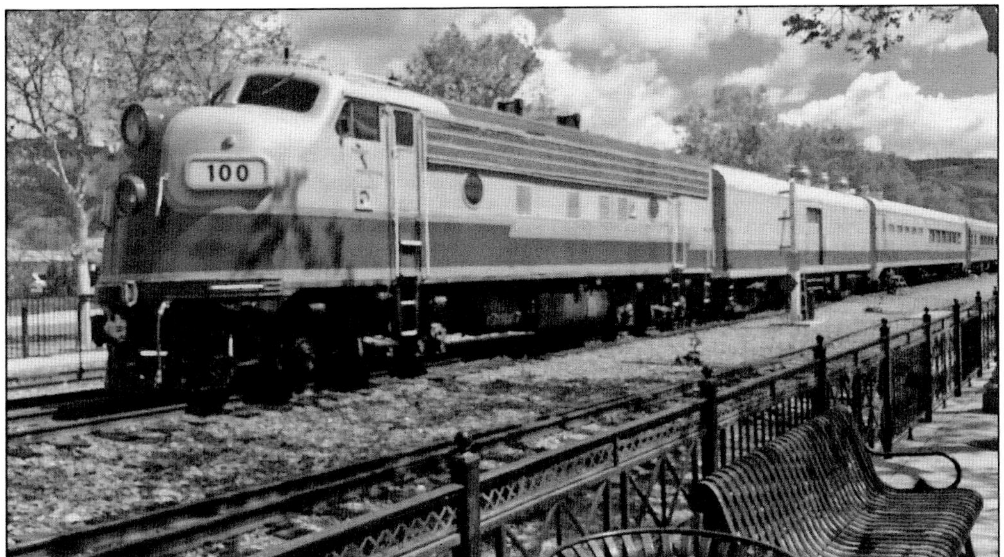

The Fillmore and Western Railway is shown as it looks today. This blue train is "dressed" just like Thomas the Tank Engine, and many families come from throughout Ventura County to ride it. The children enjoy these special excursions. Fillmore was founded because the Southern Pacific Railroad ran through the heart of the new community. The railroad transported all the fruit to Los Angeles as well. Dave Wilkinson, along with his wife, Tresa, owns and manages the railway.

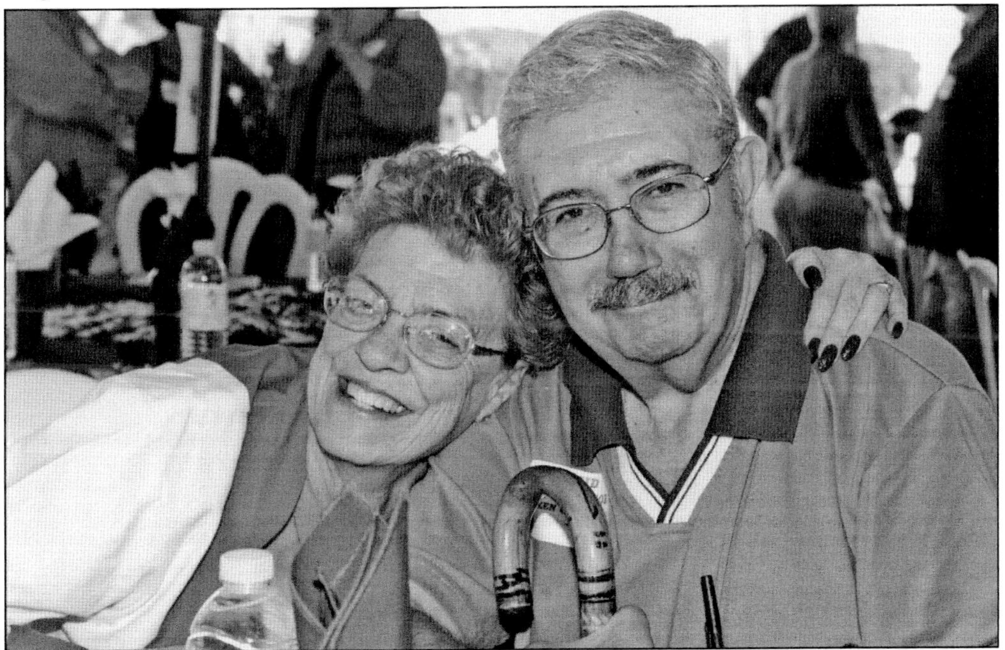

Sharon Horn Villasenor and her husband, David Villasenor, both have strong connections to Rancho Sespe and Fillmore. Sharon's parents raised their children in Rancho Sespe, and Helen Horn worked in the Rancho Sespe packinghouse. Sharon's stories about growing up in Rancho Sespe are very significant. David's father, Enrique "Herky" Villasenor, owned a labor camp in Fillmore, and he also managed all the meals for the workers and tenants. (Courtesy of Bob Cox family collection.)

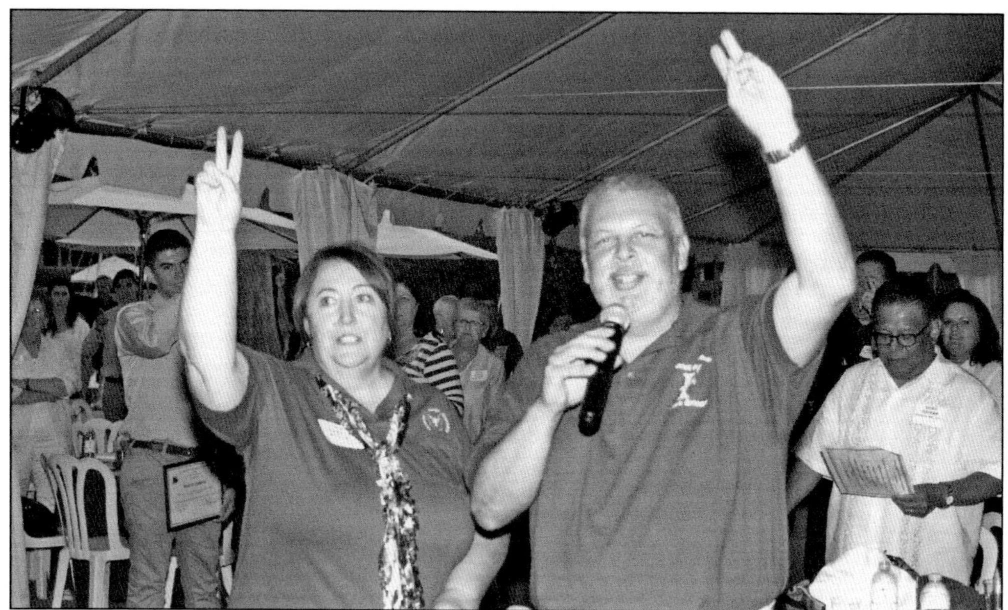
The Fillmore Alumni Association keeps everyone connected, whether they graduated from Fillmore High School or not. The group coordinates community activities for all the residents to enjoy and participate in and helps to raise scholarship money. Mark Ortega, president of the Fillmore Alumni Association, and Corinne Mozley, vice president, devote many hours to coordinating the activities and the Alumni Dinners every year. They present the scholarships to deserving Fillmore High School seniors who want to pursue a college degree. (Courtesy of Bob Cox family collection.)

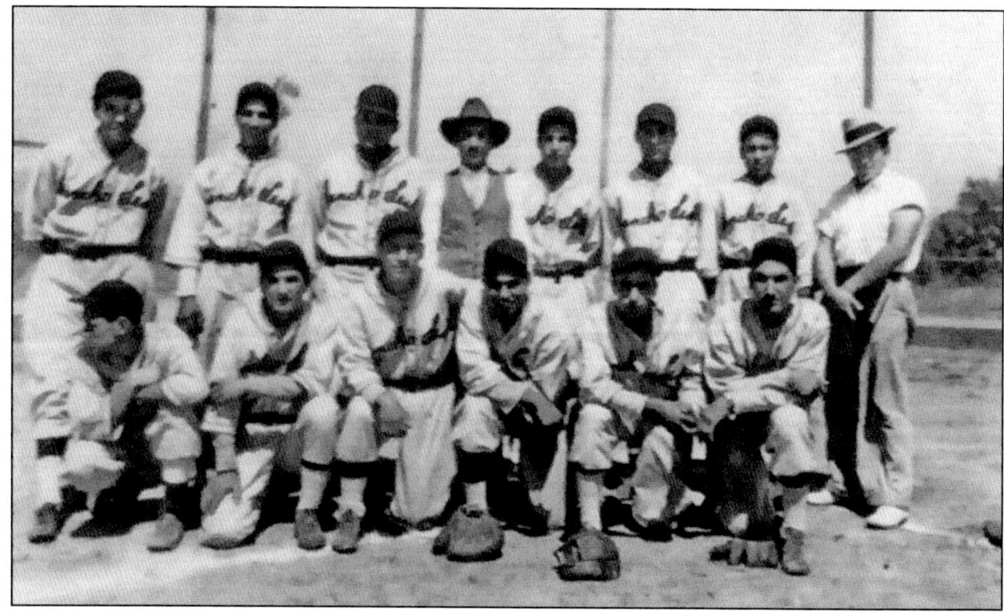
This is a photograph of the Rancho Sespe baseball team. Sports are part of the culture in Fillmore and at Rancho Sespe. The men played on Sundays, which was their day off, and after church, their families watched them play other teams throughout Ventura County. The families would take food and beverages and enjoy picnics at the baseball game.

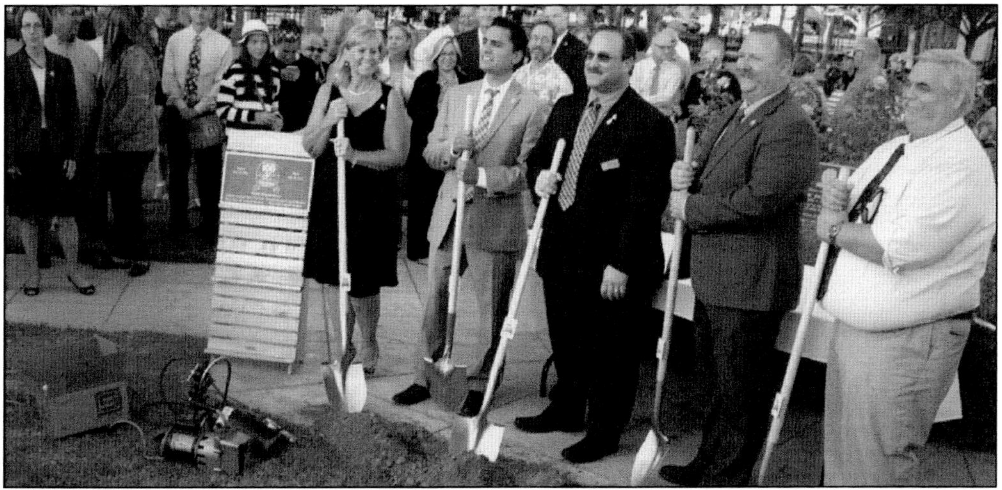

Fillmore has become progressive in that it can boast a female mayor. Diane McCall has served on the city council and is now the mayor of Fillmore. (Courtesy of *Fillmore Gazette* and publisher Martin Farrell.)

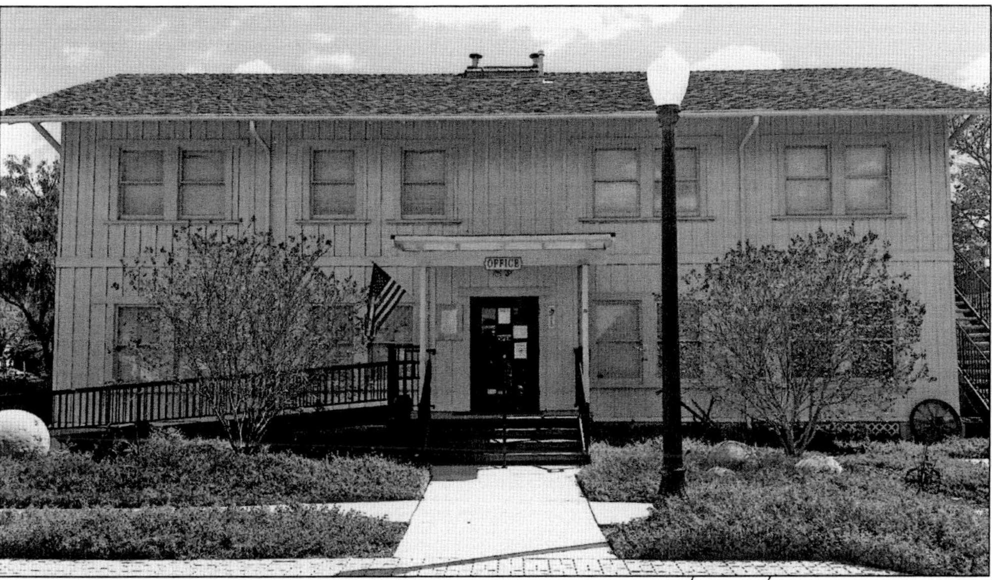

Fillmore Historical Museum, explained that she assisted Edith Jarrett in organizing the museum. The Fillmore Chamber of Commerce Board approved the idea, and donations started pouring in for displays. Dorothy came up with the idea of a museum park with three historical buildings. John Brothers, former art teacher at Fillmore High School, was commissioned to paint a mural of Fillmore's Central Avenue as it looked in the early 1900s. Dorothy then arranged to lease the land from the city for the museum, and she made all the arrangements to have the Hinckley house moved, as well as the Sespe Bunkhouse. She is proud of having assisted to open Fillmore Museum Park. The Fillmore Historical Museum is housed in the historic Sespe Bunkhouse No. 2. It is thought to be have been replicated from Bunkhouse No. 1, which was designed by Green and Green of Pasadena. Today, Martha Gentry serves as executive director of the Fillmore Historical Museum and is the resident historian. The community of Fillmore is rich in history, and many artifacts and photographs are on display in the museum for all to enjoy. (Courtesy of Ybarra family collection.)

AFTERWORD

It was summertime and there wasn't any school. Sharon Horn Villasenor remembers how she and her brother, David, climbed the roof of their house. David decided to climb on top of the roof by way of the water heater and Sharon followed him. At that moment, Mrs. Mize from next door went outside to hang out her laundry. Sharon recalls that she and David ran across to the other side of the roof and laid down because *they knew* if she saw them, she would tell Mrs. Horn. After a few minutes, they heard the noon whistle blow at the packinghouse and they had to rush to get down because their mom was going to be home soon. David scooted his way across the roof and went down by way of the water heater, but he scratched his stomach really badly (from the shingles on the roof) and Sharon decided she would not do that. She told David to run in the house and get her an umbrella. It might have worked for Mary Poppins, but it didn't work for Sharon. The umbrella inverted and she came down twice as fast. Talk about pain! Mrs. Horn arrived home and Sharon hobbled the best she could behind her mother's back and they all sat down and ate their lunch. They used to live near the Santa Clara River, which was filled with runoff water from the washing of the oranges. There was a huge oak tree across from that river that David and others used to climb. There was an orchard right next to it and there was this dwarf navel tree there; they would frequent it, pick oranges, and sit there to eat them. They were the best oranges and they were so sweet. Sharon still recalls how easy they were to peel because they were navels. She remembers growing up with four brothers was really no picnic, but they were all very good to her, especially Bob and Allen. Bob worked at the Fillmore Market and one day he took her inside the store to pick out a doll. She describes the pretty dolls standing against the wall on a shelf in the back of the store. Bob told Sharon to pick out her favorite so she selected the one next to the bride doll since she assumed it was cheaper. Of course, Bob, her brother told her to pick out her favorite since they all cost the same amount. She picked out the bride doll and she explains how she never took it out of the box and how she admired it for years. When Bob left to join the Navy, then her brother Allen took over in taking good care of Sharon and David.

Loretta Campbell Lombard remembers arriving in California from Arkansas in 1942 to visit relatives who lived in the house that everyone called "the Dip," and thus began their life on Rancho Sespe. The first house they lived in was No. 115 and she remembers her neighbor Evelyn Foster. Her mom and Loretta's mom made the girls matching dresses because Evelyn and Loretta were such good friends. The second house Loretta's family lived in was No. 73 and it was next to the "dance hall." Albert Reyes who ran the store would have big dance parties with loud but good music. One night, Loretta's mom made them lock the front door because she was afraid someone would wander inside the house from too much partying. The kindergarten was down the street and her sister Patty attended school there. Martha Stoll was the teacher. The Girl Scouts held their meetings at the kindergarten classroom and the leader was Mary Lou Pulido. Loretta remembers how all the girls loved her and how everyone learned so much from her. Loretta states she still has her Girl Scout pin to this day. The third house they lived in was No. 126 and that was next door to Lupe, Irene, and Bobby Villanueva. It was so much fun living there because Bobby's mom cooked the best Mexican food and the aroma was wonderful. They always liked being invited there to eat. The girls played jump rope, but it wasn't fun for Loretta because Lupe and Irene would go so fast and Loretta remembers she always fell and skinned her knees, yet she always went back for more. Loretta tried to jump fast, but they were faster. The Ramirez family lived on the corner and she remembers taking piano lessons from Martha Stoll and when she would go to her house for the lessons, she had to pass the Ramirez house on the way. They had two big dogs and the children were afraid of them. Loretta claims she used to dread having to walk by those dogs. They used to shout "blackouts" and they covered all the windows because of the war going on and that meant there had a blackout in the area—no lights. Many former residents have spoken about a merchant named Mike who would go to the ranch monthly. Everyone would hear him coming because he honked his distinct horn and people could hear the noise of that large truck. He had men's work pants, T-shirts and material, thread, trim, dry goods, utensils, and various useful household items. He would extend credit whenever the women needed something and could not pay him until the following month. There was a market owner in Fillmore who would bring fresh beef, vegetables, and other things not easily available at the ranch. A milkman from Golden State delivered milk, butter, and eggs. Ice was also delivered to the ranch. Mike always delivered into the community of Fillmore as well. Family members described Soledad Ybarra, author Evie Ybarra's grandmother, buying clothes and household items from Mike on a regular basis in the late 1950s when he drove his truck along Saratoga Street in the city of Fillmore. In those days, Rancho Sespe was more than a row of houses. It was a small town all its own. There was no church on the ranch, but on Saturdays, a priest from St. Francis of Assisi Catholic Church in Fillmore came and said mass in the Castorena home for all the residents. Many have fond memories of the close-knit community that was known as Rancho Sespe.

Bibliography

California Citrograph, vol. 5, May 1920 and vol. 7, 1921–1922.

Cleland, Robert G. *The Place Called Sespe: The History of a California Ranch*. Privately printed, 1940.

Fillmore Herald, March 4, 1938.

Fillmore Herald, March 11, 1938.

Perez, Vincent. *Remembering the Hacienda: History and Memory in the Mexican American Southwest*. College Station: Texas A&M University, 2006.

"Yachts of Yesteryear," August 14, 2000, www.dailypilot.com.

Ybarra, Evie. *Legendary Locals of Fillmore*. Charleston, SC: Arcadia Publishing, 2015.

Contributors

Alex Calzado
Rachel Ramirez Calzado
Cowans family
Gail Crockett
Floreine Data
Gloria Fernandez
Martha Gentry
Kenneth Glenn
Petra González
González family
Dorothy Haase
Tom Hardison
Petra González Lesperance
Becky Morales
Ernie Morales
Nancy Padelford
Henry Ramirez
Anita Robles
Esther Ramirez Robles
Rob Scherzinger
Larry Tepezano
Tomás Torres
Robert "Bobby" Villanueva
David Villasenor
Sharon Horn Villasenor

INDEX